MW00355836

Searching for ORDER IN the COMPLEXITY of Evolving Worlds

THE PUBLISHER ACKNOWLEDGES

The SFI Press would not exist without the support of William H. Miller and the Miller Omega Program.

COMPLEXITY, ENTROPY &
THE PHYSICS OF INFORMATION

*The Proceedings
of the Workshop
held May 29–June 2, 1989
in Santa Fe, New Mexico*

Volume II

WOJCIECH H. ZUREK

editor

THE SANTA FE INSTITUTE PRESS

1399 Hyde Park Road
Santa Fe, New Mexico 87501

Complexity, Entropy &
the Physics of Information, Vol. II
ISBN (HARDCOVER): 978-1-947864-31-3
Library of Congress Control Number: 2023932478

The SFI Press is made possible by the generous support
of the Miller Omega Program.

The first edition of this work was published in 1990
by Addison-Wesley Publishing Company as part of
the Santa Fe Institute Series in the Studies of Complexity.

IL N'EST PAS CERTAIN que tout soit incertain.
BLAISE PASCAL
Pensées (1669)

TRANSLATION: *It is not certain that everything is uncertain.*

VOL. II
TABLE OF CONTENTS

— Quantum Theory & Measurement —

PREFACE

Seth Lloyd, Massachusetts Institute of Technology

We hold these truths to be self-evident: that all bits are created equal; that the universe at bottom consists of information, which gives matter and energy its form; that physical dynamics transforms and processes that information. The universe computes, and this computation generates the complexity that we see around us, from the spectrum of quantum particles, to the mechanisms of life, to the structure of the cosmos.

Complexity, Entropy, and the Physics of Information is the proceedings of a workshop held by the Santa Fe Institute from May 29 to June 2, 1989. Wojciech Zurek's original introduction begins with the sentence, "The specter of information is haunting sciences," a paraphrase of the first line of the *Communist Manifesto*, Marx and Engels's call for the proletariat to throw off the chains of the bourgeoisie. Looking down on that small gathering in New Mexico from the perspective of 2022, it is difficult to remember how marginal that community of scientists was, as they tried to liberate the laws of nature from the chains imposed by the language of energy and differential equations, and to establish the language of information and computation in its place.

In the third of a century that has passed since then, however, it is not an exaggeration to say that the scientific information revolution has been accomplished, and the proletarian bit has won.

A Bit of History

1948, the hundredth anniversary of the *Communist Manifesto*, saw the publication of two momentous works, Claude Shannon's *The Mathematical Theory of Communication*, and Norbert Wiener's *Cybernetics*. Shannon showed that Ludwig Boltzmann's formula for entropy lay at the heart of technologies for communication in the presence of noise: entropy was, in fact, the amount of information required to specify the microscopic motions of atoms and molecules. Wiener's work showed the central role of communication in problems of prediction and control. Both works (in the case of Shannon, through Warren Weaver's extended introduction to the book version of the original paper) advocated the broad application of ideas of information and communication to problems in biology, the arts, and society.

The response was heady. Over the next decade, thousands of articles invoked ideas of information and communication in the service of anthropology, literature, sociology, music, etc. Hardly a field of inquiry remained untouched by Shannon's work (according to Google Scholar, Shannon's original article is the fourth most highly cited paper of all time, with the top three spots occupied by articles on biochemical measurement techniques). The deluge of half-baked works was so great that in 1956 Shannon felt called upon to write his famous *Bandwagon* article, complaining that the theory of information had "been ballooned to an importance beyond its actual accomplishments," and that theory's applications should not be "a trivial matter of translating words to a new domain, but rather the slow tedious process of hypothesis and experimental verification."

Shannon's suggestion was borne out. The original information balloon deflated. Hype gave way to hypothesis and experimental verification. The papers in *Information, Complexity, and the Physics of Information* are composed by scientists who took Shannon seriously, and who instituted a new science of physics and complexity based on information theory.

Lead-Up to the Conference

To understand the background in which the workshop took place, it's worth listing a few of the seminal scientific accomplishments leading up to the gathering in Santa Fe, which in turn planted the seeds that grew into entire fields of inquiry.

Upon arriving as a professor at MIT in 1956, Shannon established an academic discipline of coding and information theory together with Robert Fano and their student Robert Gallager. Norbert Wiener brought Warren McCulloch and Walter Pitts to MIT, where together with Jerry Lettvin they did the original work on comparing artificial neural networks with biological neural networks. More powerful digital computers were being developed, albeit slowly, and Marvin Minsky and colleagues started working on the nascent field of artificial intelligence. Because of their widespread usefulness, applications of information theory to communication and computation blossomed. A particularly elegant formulation of information theory in terms of the theory of computation is algorithmic information theory, formulated independently in the early 1960s by Ray Solomonoff, Andrey Kolmogorov, and Greg Chaitin.

Equally relevant to topics of the workshop, but less well known at the time, were direct applications of information theory to physics. In 1960, Rolf Landauer established his subsequently famous principle that erasing a bit of information required entropy to increase by $k_B \ln 2$. In 1963, Yves Lecerf showed that Turing machines could be made to be logically reversible. Lecerf's paper was only published in French, and seems not to have been known by Charlie Bennett, Ed Fredkin, Tom Toffoli, and Norm Margolus when they derived the theory of reversible computation in the 1970s and '80s. During the same period, Alexander Holevo, Lev Levitin, Carl Helstrom, and others applied information theory to quantum mechanics to derive fundamental quantum limits to physical communication channels, which are after all, at bottom, quantum mechanical.

At the same time, applications of information theory to the foundations of quantum mechanics began to appear. Bell's now-famous paper on the Einstein–Podolsky–Rosen paradox appeared in 1964 and was largely ignored for years. The idea that decoherence might provide an adequate resolution of the quantum measurement problem was introduced by H. Dieter Zeh, and subsequently explorations of the process of decoherence were explored in its environmentally induced form by Zurek *et al.* and in the form of consistent/decoherent histories by Robert Griffiths, Roland Omnès, Murray Gell-Mann, and James Hartle.

The concept of reversible computation led Paul Benioff, in 1981, to propose that computation could take place at a purely quantum-mechanical level. Benioff's work was followed by Richard Feynman's 1982 paper on quantum simulation and David Deutsch's definition of a quantum Turing machine in 1985. Quantum computation was at the time a tiny field, with fewer than a dozen papers written worldwide up to the time of the workshop (including papers by participants Asher Peres, Norm Margolus, and Wojciech Zurek).

The Workshop

The reader will have noted that a large fraction of the researchers in the nascent field of applications of information theory to problems of physics were present at the workshop. George Cowan and co-conspirators had established the Santa Fe Institute five years previously, in 1984, to investigate the sciences of complexity, and ideas of information, physics, quantum mechanics, and computation were squarely in the new Institute's sights.

The workshop was momentous for me personally: it was the first conference to which I contributed a scientific paper. *N.B.*, I had begun working on problems of information, entropy, and quantum mechanics while doing an MPhil at the University of Cambridge under the supervision of Jeremy Butterfield. When I arrived at Rockefeller University to do a PhD in 1984, I informed Nick Khuri, the department chair, of my desire to work on the topic of information theory and foundations of quantum mechanics. "Ha!" he said, "That's just for crackpots and Nobel laureates who have gone soft in the head."

I soon discovered that most of the physics world was of the same opinion, and it wasn't until 1987, when I was hard at work on my thesis (*Black Holes, Demons, and the Loss of Coherence: How Complex Systems Get Information and What They Do with It*), that I actually met more senior scientists who thought these topics of interest: first Wojciech Zurek, then Rolf Landauer and Charlie Bennett, and finally—and most spectacularly—Murray Gell-Mann, who hired me as a postdoc. On my way out to Caltech in 1988, I attended the first Santa Fe Institute summer school, which was a tremendously rich educational experience. I returned to Santa Fe in the summer of 1989 to teach in the second summer school and to attend Wojciech's workshop.

It was a tremendous honor to attend this workshop, to meet researchers whom I had only known through their seminal papers, and to begin longstanding professional and personal friendships. It remains in memory one of the most exciting conferences I have ever attended. All young researchers should have the opportunity to attend such a conference at an early stage in their careers.

The Proceedings

The workshop revolved around the question "Why?" Why quantum mechanics, and why probabilities? (Talks by Ben Schumacher, Bill Wootters, Carl Caves, Asher Peres, and Ed Jaynes.) Why complexity? (Paul Davies, Charlie Bennett, Stuart Kauffman, Tad Hogg, Jim Crutchfield, Karl Young, and myself.) There were multiple talks exploring applications of algorithmic information (Zurek, John Barrett, and others), on the emergence of the classical world from the quantum via the mechanism of decoherence (Gell-Mann and Hartle, Jonathan Halliwell, and Roland Omnès), on Hawking radiation and quantum gravity (Vladimir Mukhanov and Shin Takagi), and on alternative forms of computation (Norm Margolus, Tom Toffoli, W. G. Teich, and Günter Mahler). The corresponding articles are well worth reading today, not only for historical interest: many of the questions raised during this workshop are equally relevant today, and remain unanswered.

The workshop began with John Wheeler's famous "It from bit" talk, which he gave many times during this period. The article in the proceedings reflects Wheeler's style: comprehensive, technical, and mystical. Three years later, I hosted Wheeler's Los Alamos physics colloquium on the same topic. While he was ascending the steps to the stage, he dropped his box of transparencies, sending them flying. I offered to help him sort them back into the proper order, but he said that would not be necessary. He then gave the colloquium with his slides ordered randomly. The remarkable thing was that this version of his talk was indistinguishable from the version he gave at the workshop: the talk was permutation invariant!

The Future of 1989, as Seen from 2022

To its participants, the workshop in Santa Fe was a place to meet with other researchers to try to establish that information and computation are, fundamentally, what makes the universe work. This view was exceptionally rare amongst scientists at that time, and accordingly the talks and discussions at the workshop were particularly precious to the participants.

If we look at the themes that the participants in the workshop believed should be unified, we see that the intervening third of a century has fully justified this belief. The information-theoretic and computational theory of entropy as information now prevails. Environmentally induced decoherence and decoherent/consistent histories have been reconciled. The idea that quantum gravity should necessarily be approached using methods of quantum information and quantum computation has been adopted even by string theorists and their successors.

The most important development that took the information processing picture of the universe from margin to mainstream was quantum computing. In 1989, only half a dozen researchers worldwide had written about quantum computing, and half of those were at the workshop. At the time, no practical application of quantum computing was known: partly because of the lack of quantum algorithms, but mostly because no remotely practical way of building a quantum computer was known. Five years later, in 1994, the landscape of quantum computing altered dramatically when Peter Shor showed that quantum computers, if they could be built, would be able to factor large numbers and break public-key cryptography, thereby providing a "killer app" for quantum computers. I had already shown in 1993 that coherent, quantum computing could be performed using electromagnetic resonance (a method partly inspired by Teich and Mahler's talk at the workshop on classical molecular computers!), and Ike Chuang, Neil Gershenfeld, and I demonstrated the first quantum algorithms experimentally a few years later.

Over the next decade, the rapidly increasing interest in quantum computation took the tiny, marginal fields of quantum information and quantum foundations and "institutionalized" them as part of the standard canon of physics. The last decade has seen large investments in quantum information and quantum computing both by national governments and by private companies, and the number of researchers in the field has grown from half a dozen to thousands. It is a wonderful honor for the field that while this preface was being written, the 2022 Nobel Prize in Physics was awarded to John Clauser, Alain Aspect, and Anton Zeilinger for experimental demonstrations of the counterintuitive aspects of quantum entanglement, an award that would have excited astonishment and applause from the 1989 participants.

The Future

Just because the world view of the workshop—that the universe consists, at bottom, of information in its quantum form—has been vindicated and adopted by the broader scientific community does not mean that all the problems raised then have been solved. Far from it. Quantum information helps us to understand gravity, but gravity has not been fully quantized. Decoherence helps us to understand the quantum measurement problem, but the measurement problem remains. Small-scale quantum computers have been constructed, but building large-scale quantum computers remains a daunting technological task. Nonetheless, as long as we can continue to explore and to establish the sciences of information, computation, and complexity, the governance of the bit, by the bit, and for the bit shall not perish from this universe. ⸎

COMPLEXITY, ENTROPY, AND THE PHYSICS OF INFORMATION— A MANIFESTO

Wojciech H. Zurek, Los Alamos National Laboratory and the Santa Fe Institute

The specter of information is haunting sciences. Thermodynamics, much of the foundation of statistical mechanics, the quantum theory of measurement, the physics of computation, and many of the issues of the theory of dynamical systems, molecular biology, genetics, and computer science share information as a common theme. Among the better established, but still mysterious, hints about the role of information are:

- *A deep analogy exists between thermodynamic entropy and Shannon's information-theoretic entropy.* Since the introduction of Maxwell's demon and, particularly, since the celebrated paper of Szilard and even earlier discussions of Smoluchowski, the operational equivalence of the gain of information and the decrease of entropy has been widely appreciated. Yet, the notion that a subjective quantity such as information could influence "objective" thermodynamic properties of the system remains highly controversial.

It is, however, difficult to deny that the process of information gain can be directly tied to the ability to extract useful work. Thus, questions concerning thermodynamics, the second law, and the arrow of time have become intertwined with a half-century-old puzzle, that of the problem of measurements in quantum physics.

- *Quantum measurements* are usually analyzed in abstract terms of wave functions and Hamiltonians. Only very few discussions of the measurement problem in quantum theory make an explicit effort to consider the crucial issue— the transfer of information. Yet obtaining knowledge is the very reason for making a measurement. Formulating quantum measurements and, more generally, quantum phenomena in terms of information should throw a new light on the problem of measurement, which has become difficult to ignore in light of new experiments on quantum behavior in macroscopic systems.

The distinction between *what is* and *what is known to be*, so clear in classical physics, is blurred, and perhaps does not exist at all on a quantum level. For instance, energetically insignificant interactions of an object with its quantum environment suffice to destroy its quantum nature. It is as if the "watchful eye" of the environment "monitoring" the state of the quantum system forced it to behave in an effectively classical manner. Yet, even phenomena involving gravity, which happen on the most macroscopic of all the scales, bear the imprint of quantum mechanics.

In fact it was recently suggested that the whole universe— including configurations of its gravitational field—may and should be described by means of quantum theory. Interpreting results of the calculations performed on such a "wave function of the universe" is difficult, as the rules of thumb usually involved in discussions of experiments on atoms, photons, and electrons assume that the "measuring apparatus" as well as "the observer" are much larger than the quantum system. This is clearly not the case when the quantum system is the whole universe. Moreover, the transition from quantum to classical in the early epochs of the existence of the universe is likely to have influenced its present appearance.

- *Black hole thermodynamics* has established a deep and still largely mysterious connection between general relativity, quantum mechanics, and statistical mechanics. Related questions about the information capacity of physical systems, fundamental limits on the capacity of communication channels, the origin of entropy in the universe, etc. are subjects of much recent research.

- xxiii -

The three subjects above lie largely in the domain of physics. The following issues forge connections between the natural sciences and the science of computation or, rather, the subject of information processing regarded in the broadest sense of the word.

- *Physics of computation* explores limitations imposed by the laws of physics on the processing of information. It is now established that both classical and quantum systems can be used to perform computations *reversibly*. That is, computation can be "undone" by running the computer backwards. It appears also conceivable that approximately reversible computer "chips" can be realized in practice. These results are of fundamental importance, as they demonstrate that, at least in principle, processing of information can be accomplished at no thermodynamic cost. Moreover, such considerations lead one to recognize that it is actually the erasure of the information which results in the increase of entropy.

The information which is being processed by the computer is a concrete "record," a definite sequence of symbols. Its information content cannot be represented adequately in terms of Shannon's probabilistic definition of information. One must instead quantify the information content of the specific, well-known "record" in the memory of the computer—and not its probability or frequency of occurrence, as Shannon's formalism would demand. Fortunately, a relatively new development—a novel formulation of information theory—has already been put forward.

- *Algorithmic randomness*—an alternative definition of the information content of an object based on the theory of computation rather than on probabilities—was introduced more than two decades ago by Solomonoff, Kolmogorov, and Chaitin. It is equal to the size of the shortest message which describes this object. For instance, the string of 10^5 0s and 1s

$$01010101010101\ldots$$

can be concisely described as "$5 \cdot 10^4$ 01 pairs." By contrast, no concise description can be found for a typical, equally long string of 0s and 1s generated by flipping a coin. To make this definition more rigorous, the universal Turing machine —a "generic" universal computer—is usually considered to be the "addressee" of the message. The size of the message is then equal to the length—in number of bits—of the shortest program that can generate a sufficiently detailed description (for example, a plot) of the object in question.

It is tempting to suggest that physical entropy—the quantity which determines how much work can be extracted from a physical system—should take into account the system's algorithmic randomness. This suggestion can be substantiated by detailed discussions of examples of computer-operated engines as well as by results concerning the evolution of entropy and algorithmic randomness in the course of measurements. It provides a direct link between thermodynamics, measurements, and the theory of computation. Moreover, it is relevant to the definition of complexity.

- *Complexity*, its meaning, its measures, its relation to entropy and information, and its role in physical, biological, computational, and other contexts have become subjects of active research in the past few years.

Proceedings

This book emerged from a meeting held during the week of May 29 to June 2, 1989, at St. John's College in Santa Fe under the auspices of the Santa Fe Institute. The (approximately forty) official participants, as well as equally numerous "groupies," were enticed to Santa Fe by the above "manifesto." The book—like the "Complexity, Entropy, and the Physics of Information" meeting—not only explores the connections between quantum and classical physics, information and its transfer, computation, and their significance for the formulation of physical theories, but also considers the origins and evolution of the information-processing entities, their complexity, and the manner in which they analyze their perceptions to form models of the universe. As a result, the contributions can be divided into distinct sections only with some difficulty.

Indeed, I regard this degree of overlap as a measure of the success of the meeting. It signifies consensus about the important questions and on the anticipated answers: they presumably lie somewhere in the "border territory" where information, physics, complexity, quantum, and computation all meet. ⚓

Acknowledgments

I would like to thank the staff of the Santa Fe Institute for excellent (and friendly) organizational support. In particular, Ginger Richardson was principally responsible for letting "the order emerge out of chaos" during the meeting. And somehow Ronda Butler-Villa managed the same feat with this volume.

I would like to gratefully acknowledge the Santa Fe Institute, the Air Force Office for Scientific Research, and the Center for Nonlinear Studies, Los Alamos National Laboratory, for the financial (and moral) support which made this meeting possible.

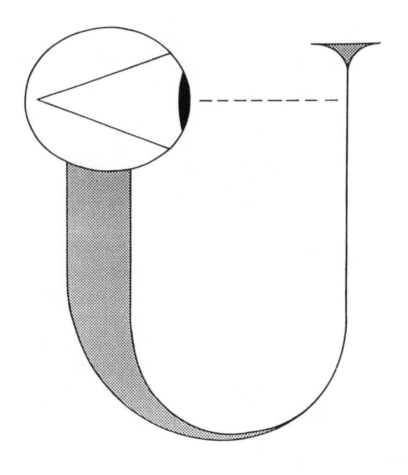

PHYSICS OF
COMPUTATION

ojɲ

PARALLEL QUANTUM COMPUTATION

Norman Margolus, MIT Laboratory for Computer Science

Results of Feynman and others have shown that the quantum — 395 —
formalism permits a closed, microscopic, and locally interacting system
to perform deterministic serial computation. In this paper we show that
this formalism can also describe deterministic parallel computation.
Achieving full parallelism in more than one dimension remains an
open problem.

Introduction

In order to address questions about quantum limits on computation, and the possibility of interpreting microscopic physical processes in informational terms, it would be useful to have a model which acts as a bridge between microscopic physics and computer science.

Feynman (1985) and others (Benioff 1986; Margolus 1986) have provided models in which closed, locally interacting microscopic systems described in terms of the quantum formalism perform deterministic computations. Up until now, however, all such models implemented deterministic *serial* computation, i.e., only one part of the deterministic system is active at a time.

We have the prejudice that things happen everywhere in the world at once, and not sequentially like the raster scan which sweeps out a television picture. It would be surprising, and perhaps a serious blow to attempts to ascribe some deep significance to information in physics, if it were impossible to describe *parallel* computations within the quantum formalism.

In this paper, we extend the discussion of a previous paper (Margolus 1986) to obtain for the first time a satisfactory model of parallel "quantum" computation, but only in one dimension. The two-dimensional system discussed in Margolus (1986) is also shown to be a satisfactory model, but the technique used here only allows one dimension to operate in parallel: the more general problem of the possibility of fully parallel two- or three-dimensional quantum computation remains open.

Computation

The word *computation* is used in many contexts. Adding up a list of numbers is a kind of computation, but this task requires only an adding machine, not a general-purpose computer. Similarly, we can *compute* the characteristics of airflow past an aircraft's wing by using a wind tunnel, but such a machine is no good for adding up a list of numbers.

An adding machine and a wind tunnel are both examples of computing machines: machines whose real purpose is not to move paper or air, but to manipulate information in a controlled manner. It is the rules that transform the information that are important: whether the adding machine uses enormous gears and springs or microscopic electronic circuits, as long as it follows the addition algorithm correctly, it is acting as an adding machine.

A *universal computer* is the king of computing machines: it can simulate the information transformation rules of any physical mechanism for which these rules are known. In particular, it can simulate the operation of any other universal computer—thus all universal computers are equivalent in their simulation capabilities. It is an unproven, but thus far uncontradicted contention of computer theory that no mechanism is any more universal than a universal digital computer, i.e., one that manipulates information in a discrete form.

Assuming a finite universe, no machine can have a truly unbounded memory; what we mean when we talk about a *general-purpose computer* is a machine that, if it could be given an unbounded amount of memory, would be a universal computer. (In common usage, the terms *general-purpose computer* and *computer* are synonymous.) Similarly, when we talk about a finite set of logic elements as being *universal,* we mean that an unbounded collection of such elements could constitute a universal computer.

An adding machine is not a general-purpose computer: a certain minimum level of complexity is required before universal behavior is possible. This complexity threshold is quite low: aside from memory, a few dozen logical NAND gates, suitably connected, can be a computer. On the other hand, some modern computers contain millions of logic elements in their central processors: this doesn't let these computers solve any problems that the humblest microcomputer couldn't solve; it simply lets them run faster. Except for speed and memory capacity, there is no difference in computational capability between a Cray-XMP and an IBM-PC.

Quantum Computation

Although all general-purpose computers can perform the same computations, some of them work faster, use less energy, weigh less, are quieter, etc., than others. In general, some make better use of the computational opportunities and resources offered by the laws of physics than do others. For example, since signals travel so slowly (it takes about a nanosecond to go a foot at

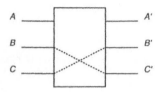

Figure 1. Fredkin gate. The top *control* input goes through unchanged ($A' = A$), and the bottom inputs either go straight through also (if $A = 1$) or cross (if $A = 0$, then $B' = C$ and $C' = B$).

the speed of light), there is a tremendous speed advantage in building computers which have short signal paths. Modern microprocessors have features that are only a few hundred atoms across: such small components can be crowded close together, allowing the processor to be small, light, and fast.

As we try to map our computations more and more efficiently onto the laws and resources offered by nature, we are eventually confronted with the question of whether or not we can arrange for extremely microscopic physical systems to perform computations. What we ask here is in a sense the opposite of the hidden-variables question: we ask not whether a classical system can simulate a quantum system in a microscopic and local manner, but rather whether a quantum system can simulate a classical system in such a manner.

All of our discussion of quantum computation will be based on autonomous systems: we prepare the initial state and let the system undergo a Schrödinger evolution as an isolated system, and after some amount of time we examine the result.[1] Since the Schrödinger evolution is unitary, and hence invertible, we must base our computations on reversible logic (Fredkin and Toffoli 1982).

---

[1] For some types of computation, we can't set a very good limit on how long we should let it run before looking. In such cases, we would simply start a new computation if we look and find that we aren't finished.

Reversible Computation

Until recently, it was thought that computation is necessarily irreversible: it was hard, for instance, to imagine a useful computer in which one could not simply erase the contents of a register. It was to most people a rather surprising result[2] that computers can be constructed completely out of invertible logic elements, and that such machines can be about as easy to use as conventional computers. This result has thermodynamic consequences, since it turns out that a reversible computer is the most energy-efficient engine for transforming information from one form to another. This result also means that computation is not necessarily a (statistically irreversible) macroscopic process.

~399~

As an example of an invertible logic element, consider the *Fredkin gate* of figure 1. This gate is in fact its own inverse (two connected in series give the identity function), and it is a universal logic element: you can construct any invertible logic function out of Fredkin gates. A logic circuit made out of Fredkin gates looks much like any conventional logic circuit, except that special "mirror-image circuit" techniques are used to avoid the accumulation of undesired intermediate results that we aren't allowed to simply erase (see Fredkin and Toffoli [1982] for more details).

Feynman made a quantum system simulate a collection of invertible logic gates connected together in a combinational circuit (i.e., one without any feedback).[3] In Feynman's construction, only one logic element was active (i.e., transforming its

---

[2] See Bennett (1973), Fredkin and Toffoli (1982), Landauer (1961), Margolus (1984), and Toffoli (1977).

[3] Although combinational circuitry can perform any desired logical function, computers are usually constructed to run in a cycle, reusing the same circuitry over and over again. The parallel models discussed later in this paper run in a cycle.

inputs into outputs) at any given time: the different gates were activated one at a time as they were needed to act on the output of gates that were active earlier. We can imagine a sort of "fuse" running through our circuit: as the active part of the fuse passes each circuit element in turn, it activates that element. Using a collection of two-state systems (which he called *atoms*) to represent bits, Feynman made a "quantum" version of this model. In what follows, we will think of our two-state systems as spin-$\frac{1}{2}$ particles.

Feynman's Quantum Computer

In 1985, Richard Feynman (1985) presented a model of computation which was quantum-mechanically plausible: there seems to be no fundamental reason why a system like the one he described couldn't be built.[4] In his idealization, he managed to arrange for all of the quantum uncertainty in his computation to be concentrated in the time taken for the computation to be completed, rather than in the correctness of the answer. Thus if his system is examined and a certain bit (state of a spin) indicates that the computation is done, then the answer contained elsewhere in the system is always correct. What's more, he managed to make his computation run at a constant rate.

His system consists of two parts: a collection of reversible logic gates, each made up of several interacting spins, and a chain of "clock" spins which passes next to each gate in turn. Note that we will think of each wire that runs between two gates as being a very simple reversible gate: one that exchanges the values of the spins at its two ends. In this way we are able to write down a unitary operator F_k that describes the

---

[4]Less-physical models were proposed earlier by Benioff (1982), who seems to have been the first to raise the question of quantum computation in print.

desired behavior of the kth gate: for a given invertible gate such as the Fredkin gate or a wire, we can write this operator down explicitly in terms of raising and lowering operators. For example, for a wire F_i joining spin a of gate m and spin b of gate n, the rule is

$$F_i = a_m b_n^\dagger + a_m^\dagger b_n + a_m a_m^\dagger b_n b_n^\dagger + a_m^\dagger a_m b_n^\dagger b_n,$$

where a and b are lowering operators at the two spins and $a^\dagger$ and   ~401~
$b^\dagger$ are their Hermitian adjoints, which are raising operators on the two spins.

    Without any claim yet to a connection with quantum mechanics, we can cast the overall logical function implemented by an N-gate invertible combinational logic function into the language of linear operators acting on a tensor product space as follows:

$$F = \sum_{k=1}^{N} F_k c_k c_{k+1}^\dagger \tag{1}$$

where c_k is the lowering operator on the clock spin that passes next to the kth gate F_k. If we start all of the clock spins off in the down state, except for the spin next to the first gate, then if F acts on this system, only the term

$$F_1 c_1 c_2^\dagger$$

will be non-vanishing. This term will cause the spins acted upon by the first gate to be updated: the first clock spin will be turned down, and the second clock spin will go up. Similarly, if F acts again, the second gate will update, and the up spin will move to the third position. Clearly, if the initial state has only a single clock spin up, F will preserve that property. Using the position of the up clock spin to label the state, if $|1\rangle$ is the initial state, then $F|1\rangle = |2\rangle$, and in general $F|k\rangle = |k + 1\rangle$. We have thus been able to write the forward time step operator as a sum of local pieces by serializing the computation—only one gate in the circuit is active during any given step.

Notice that the operator $F_k^\dagger$ is the inverse of F_k, since the role of raising and lowering operators is interchanged. Similarly, $F^\dagger$ is the inverse of F, since each term of the former undoes the action of the corresponding term of the latter, including moving the clock spin back one position. Now if we add together the forward and backward operators, we get an Hermitian operator $H = F + F^\dagger$, which is the sum of local pieces, each piece acting only on a small collection of neighboring spins (a gate). At this point we make contact with quantum mechanics, by seeing what happens if we use this H as the Hamiltonian in a Schrödinger evolution.

If we expand the time evolution operator $U(t) = e^{-iHt}$, we get

$$U(t) = 1 - iHt - \frac{H^2 t^2}{2} + \cdots$$
$$= 1 - i(F + F^\dagger)t - \frac{(F + F^\dagger)^2 t^2}{2} + \cdots,$$

and so we get a sum of terms, each of which is proportional to F or $F^\dagger$ to some power. Thus if $|k\rangle$ is evolved for a time t, it becomes $e^{-iHt}|k\rangle$, which is a superposition of configurations of the serialized computation that are legitimate successors and predecessors of $|k\rangle$: each term in the superposition has a single clock spin at some position, and the computation is in the corresponding state.

Feynman now noted that the operators F_k don't affect the dynamics of the c_ks: we can consider $F = \sum_{k=1}^N c_k c_{k+1}^\dagger$ for the purposes of analyzing the evolution of the clock spins. But then $H = F + F^\dagger$ supports superpositions of the one-spin-up states called spin waves, as is well known. When we add back in the F_ks, the computation simply rides along at a uniform rate on top of the clock spin waves. This point will be discussed in more detail below, when we extend this serial model to deal with parallel computation.

Parallel Computation

Serial computers follow an algorithm step by step, completing one step before beginning the next; parallel computers make it possible to do several parts of the problem at once in order to finish a computation sooner. Although Feynman's construction is based on a serial model, his idea of concentrating all of the quantum uncertainty into the time of completion, while leaving none in the correctness of the computation, can be extended to parallel computations (Margolus 1986). Maintaining correctness is again achieved simply by construction of the Hamiltonian: states in the Hilbert space that correspond to configurations on a given computational orbit form an invariant subspace under the Schrödinger evolution. This property of the Hamiltonian does not, in general, say anything about the rate at which we can compute. Here we show that Feynman's technique for making a serial model of quantum computation run at a constant rate can, in fact, also be extended to apply to a parallel system, in particular to the one-dimensional analog of the case considered in Margolus (1986). From this, we can derive a way of making the two-dimensional system in Margolus (1986) compute at a constant rate, but with parallelism that extends over only one dimension.

For simplicity, our discussion of parallel computers will be confined to cellular automata (CA): uniform arrays of computing elements, each connected only to its neighbors. These systems can be universal in the strong sense that a given universal cellular automaton (assuming it's big enough)

~403~

can simulate any other computing structure of the same dimensionality at a rate that is independent of the size of the structure.[5] By showing that, given any desired (synchronous) CA evolution, we can write down a Hamiltonian that simulates it, we will have shown that the quantum-mechanical formalism is computationally universal in this strong sense, at least for one-dimensional rules.

Feynman's model involved only states in which a single site was active at a time. In order to accommodate both neighbor interactions and parallelism in quantum mechanics, we find that we are forced to consider asynchronous (no global time) computing schemes (but still employing invertible logic elements). For suppose that our Hamiltonian is a sum of pieces each of which only involves neighbor interactions:

$$H = \sum_{x,y,z} H_{x,y,z}. \qquad (1)$$

Then consider the time evolution $1 - iHt$ over an infinitesimal time interval. When this operator acts on a configuration state of our system, we get a superposition of configuration states: one term in the superposition for every term in the sum (eq. 1) above. If we want all of the terms in this superposition to be valid computational states, then we must allow configurations in which one part has been updated while everything else has been left unchanged.

Local Synchronization

One can perform an effectively synchronous computation using an asynchronous mechanism by adding extra state variables to keep track of relative synchronization (how many

---

[5]This isn't the usual definition of universality in CA, but it is the one that we'll use here.

more times one portion of the system has been updated than an adjacent portion). To use an analogy, consider a bucket brigade carrying a pile of stones up a hill. You hand a stone to the first person in line, who passes it on to the next, and so on up the hill. An *asynchronous* computation would correspond to every individual watching the person ahead of himself, and passing his stone along when the next person has gotten rid of theirs. This involves only local synchronization. A *synchronous* computation would correspond to having everyone pass on their stones whenever they hear the loud tick of a central clock. Notice that both schemes get exactly the same sequence of stones up the hill; only the timing of when a given stone moves from hand to hand changes.

Now let us consider a one-dimensional cellular automaton. We imagine a row of cells, each containing a few bits of state. Our evolution will consist of two phases: first, we group the cell at each even-numbered position with the cell to its right, and perform a logical transformation on the state of these two cells; then we regroup the cells so that each even-numbered cell is associated with the cell to its left, and again we update the pair. We alternate these two kinds of steps to produce a dynamics. Notice that if the transformation we perform on each pair of cells is an invertible logic function, then the overall dynamics will be invertible.

Since cells are updated in pairs, it is really unnecessary for the entire system to be globally synchronous: we can achieve effectively the same result by local means. Imagine that we take our configuration of cells, and to each cell we add an extra number, which is the number of times that cell has been updated. In a synchronous updating scheme, all cells would start out with this number set to zero, and this number would increment uniformly throughout the system: if one cell is at step 27, all cells are. But suppose we start out with the same initial data, and only update one pair (with the appropriate grouping for an even-numbered step). Since the result of this updating only depends on the contents of these two cells, it makes no difference whether or not any other cells have been updated yet. Next, we could update some more pairs. Now suppose two adjacent pairs have been updated: we have four consecutive cells that correspond to the synchronous time step number one, and are labeled as having been updated once. The middle two cells of these four are a correct group for an odd-numbered synchronous step, and so we can update this odd pair and label them as having been updated twice. Each of these two cells is now ready to be updated again as part of even pairs, as soon as the adjacent cells catch up! Thus we can perform an asynchronous updating of pairs, using the count of updates for each cell to tell us when adjacent cells can be updated as a pair. As long as we observe this protocol, we can update cells in any order and retain the property that any cell that is labeled as having been updated n times is at the same state that it would have been if the whole system had been updated synchronously n times.

Figure 2. A section of a one-dimensional pairing automaton showing only the states of the clock bits in each cell. The solid bars bracket the pairing used for even times, the dotted for odd times.

Notice that with this scheme, two adjacent cells cannot get more than one step apart in update-count: since this count is only used to tell whether a given cell is using the even step pairing or the odd step pairing, and to tell if adjacent cells are at the same step, we only need to look at the least significant bit of the update-count. Thus if we take our original synchronous automaton and add a single bit of update-count to each cell, we can run the system asynchronously while retaining a perfectly synchronous causality.

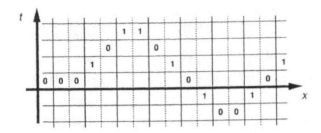

Figure 3. A spacetime diagram showing relative times of adjacent clock spins corresponding to the data in the previous figure. Pairing of cells is indicated as before.

In figure 2 we show a possible state for the update-count bits (henceforth we'll call them *clock bits*) in a one-dimensional pairing automaton of the type we've been discussing, which is consistent with an evolution starting from a synchronous initial state. In figure 3 we use a spacetime diagram to integrate the relative time phases: arbitrarily calling the time at the left-hand position $t = 0$, we mark cells using the relative time information encoded in the clock bits. As we move across, if a cell is at the same time as its neighbor to the left, we mark it at the same time on this diagram; if it is ahead, we mark it one position ahead, etc. The result is a diagram illustrating the hills and valleys of time present in this configuration. Note that we can tell if a given cell in figure 2 which is at a different time phase than its neighbor to the left is ahead or behind this neighbor by seeing whether or not it is waiting for the neighbor to catch up in order to be paired with it.

Note that if we allow backward steps, this synchronization scheme still works fine: we can imagine that a backward step is simply undoing a forward step, getting us to a configuration we could have gotten to by starting at an earlier initial synchronous step and running forward.

These configurations then, with their hills and valleys of time, will be the classical configurations which our quantum system will simulate.

A "Quantum" Parallel Automaton

Again we imagine a collection of interacting spins as our computational system. Let $|n, \alpha\rangle$ be a state on our locally synchronized computational trajectory, where n refers to time and α refers to other information needed to uniquely specify a configuration. Since our configurations have no global moment of time, we use an integrated notion of time: we simply add up the equivalent synchronous times for all cells in the automaton, and divide by the number of cells in a single block. With this normalization, if we have a configuration at integrated time n and we take a step forward at a single block, then the resulting configuration will be at time $n + 1$.

We imagine that our system has two kinds of spins at each site in our one-dimensional chain of cells: data spins and clock spins. We'll let D_i be our rule for updating the block of data spins belonging to two adjacent cells at locations i and $i + 1$; $D_i^\dagger$ is the inverse rule. We imagine that we have a single spin-$\frac{1}{2}$ clock spin at each cell, and that c_i is the lowering operator acting on the spin at cell i. Now we can define F, our forward time step operator, as

$$F = \sum_{i \text{ even}} D_i c_i^\dagger c_{i+1}^\dagger + \sum_{i \text{ odd}} D_i c_i c_{i+1} = \sum_i F_i. \qquad (3)$$

This operator, acting on a state $|n, \alpha\rangle$, produces a superposition of states each of which belongs to time $n + 1$. Similarly, $F^\dagger$ takes us backward one time step. Note, however, that $F^\dagger$ is not the inverse of F. Nevertheless, on the subspace of computational configurations (those that can be obtained by a sequence of local updates starting from a synchronous configuration), F and $F^\dagger$ commute: this property, which will be proven below, will be crucial in our construction.

As before, we let $H = F + F^\dagger$, and if we expand the time evolution operator $U(t) = e^{-iHt}$ we get a superposition of terms, each involving products of F_i and $F_j^\dagger$ for various i and j.

Since each such term, acting on a computational configuration, gives us another computational configuration (by construction of the clock bits), the time evolution U doesn't take us out of our computational subspace.

RUNNING IN PARALLEL

Now we would like to have our parallel computation run forward at a uniform rate. We are imagining that our space is periodic: the chain of cells is finite and the ends are joined. Designating one particular state of the equivalent globally synchronous computation as $t = 0$, we can assign a value of t to every configuration on each synchronous computational orbit, and from these assign a value of n to the integrated time on every locally synchronized computational configuration. Thus we can construct an operator N which, acting on a configuration $|n, \alpha\rangle$, returns n:

$$N |n, \alpha\rangle = n |n, \alpha\rangle.$$

From this we can construct a *computational velocity* operator V:

$$V = \frac{[N, H]}{i} = \frac{[N, F]}{i} + \frac{[N, F^\dagger]}{i}.$$

But $NF |n, \alpha\rangle = (n + 1)F |n, \alpha\rangle$, since F takes $|n, \alpha\rangle$ into a superposition of states all of which correspond to time $n + 1$, and so

$$[N, F] |n, \alpha\rangle = (n + 1)F |n, \alpha\rangle - nF |n, \alpha\rangle = F |n, \alpha\rangle$$

and, similarly, $[N, F^\dagger] |n, \alpha\rangle = -F^\dagger |n, \alpha\rangle$. Thus on this subspace,

$$V = \frac{F - F^\dagger}{i}.$$

Now for the average computational velocity $\langle V \rangle = d\langle N \rangle/dt$ to be constant, we would like V to commute with H. So the question becomes, "does V commute with H?" Now $[V, H] =$

$[(F - F^\dagger)/i,\ F + F^\dagger] = 2[F, F^\dagger]/i$, and so this is the same as the question "does F commute with $F^\dagger$?"

Each term in the product $F\,F^\dagger$ involves one F_j and one $F_k^\dagger$. Clearly if $|j - k| \geq 2$, then $[F_j, F_k^\dagger] = 0$, since the two operators act on disjoint sets of spins. If $|j - k| = 1$, then the product $F_j F_k^\dagger$ vanishes when applied to a computational state, since either F_j or $F_k^\dagger$ vanishes: either the pair of cells at k and $k + 1$ are not ready to take a step backward (and so $F_k^\dagger$ vanishes), or if they are ready to go back and $F_k^\dagger$ acts on them, then in the resulting configuration these two cells are only ready to take a step forward if they are paired together again, and so F_j vanishes. Thus the commutator of F and $F^\dagger$ can be written

$$[F, F^\dagger] = \sum_k [F_k, F_k^\dagger] = \sum_k F_k F_k^\dagger - \sum_k F_k^\dagger F_k,$$

which, when applied to a computational configuration, just gives the difference between the number of blocks that are ready to go backward and the number that are ready to go forward. Now the question of commutation is reduced to a question about the computational configurations: "Is it true that the number of blocks ready to go forward is always equal to the number ready to go back?"

For the two-dimensional case considered in Margolus (1986) the answer is no, but for a one-dimensional automaton with periodic boundaries the answer is yes: in a flat (globally synchronous) configuration, the answer is clearly yes, and it is easy to check that any sequence of updates preserves this property.[6]

Now we can make our cellular automaton run at a uniform rate: we use as our initial state a superposition of eigenstates of V

---

[6]Equivalently, one can simply observe that between every two blocks that are ready to go forward there is always a block that is ready to go back, and vice versa, and so in a periodic configuration the number ready to go forward is always equal to the number ready to go back.

which has a fairly narrow ΔN, so that the integrated time in our computation is fairly definite.[7] Since $\langle V \rangle$ is constant, this state will evolve at a uniform rate, as desired.

RELATING THE MODELS

It turns out that the one-dimensional version of Feynman's serial model is a special case of the model discussed above: if we complement the meaning of every second clock spin (say, all the ones at even positions), equation 3 becomes

$$F = \sum_{i \text{ even}} D_i c_i c_{i+1}^{\dagger} + \sum_{i \text{ odd}} D_i c_i c_{i+1}^{\dagger} = \sum_{i} D_i c_i c_{i+1}^{\dagger},$$

which is of exactly the same form as equation 1. An initial state containing a single *up* clock spin and all the rest down would correspond, in our parallel system of equation 3, to all of the even clock spins up and all of the odd ones down, except for the spin at the active position k, which is the same as its two neighbors. Since updating in our parallel model only occurs at positions where two adjacent clock spins are the same, there are only two active blocks in such an alternating configuration: the block involving k and $k + 1$, which will be a step forward if updated, and the block involving k and $k - 1$, which will be a step back if updated. If we draw a spacetime diagram of the clock spins around position k (see fig. 4), showing the relative synchronization implied by the alternating pattern of clock spins, we see that it forms a staircase with a landing that moves up and down in time as its leading edge or trailing edge is updated. Because the space is periodic, the top of this staircase is connected to the bottom: this configuration is not on the orbit of any synchronous parallel computation.

---

[7] This also avoids the necessity of performing the whole computation ahead of time in order to construct the initial superposition: we simply truncate the small-amplitude long-time terms of our initial superposition, effectively adding a small error term to our state whose amplitude doesn't grow with time (Zurek 1984).

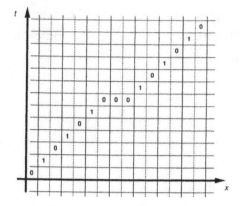

Figure 4. A spacetime diagram of the active region of a parallel one-dimensional cellular automaton with a staircase configuration.

Fixing the Two-Dimensional Model

In Margolus (1986) I gave a two-dimensional analog of the parallel model discussed here, using a particular universal reversible cellular automaton which was updated using a two-dimensional version of the locally synchronized block partitioning discussed above. There I was unable to make the model run at a uniform rate; the parallel technique used above can in fact be extended to make this earlier model run, but with only one dimension of parallelism. The idea is to sweep a one-dimensional parallel active region across the two-dimensional system using staircase and landing configurations analogous to what we saw in the previous section: we initialize the rows of our two-dimensional clock spins (they were called guard bits in Margolus [1986]) with an alternating pattern of horizontal stripes (all of the even rows up and the odd rows down) except for a single row (the active region) that is the same as its two neighboring rows. Then every column contains exactly one segment with three consecutive clock spins that are the same, and in fact each column of clock spins, when represented on a spacetime diagram, looks exactly like figure 4. It is easy to verify that this property is preserved by the dynamics, and that the dynamics of the active region is isomorphic to that of our one-dimensional parallel model. Thus if we make a wave packet state out of configurations on the same computational orbit as a staircase with a landing, we can make this wave packet run repeatedly across our system, doing a line of updates in parallel as it travels up the staircase.

Note that the CA model in Margolus (1986) has the property that one can perform any desired computation by constructing patterns of up and down values in the data spins that resemble conventional computer circuitry: gates, signals, wires, etc. In such patterns, no signals need ever go outside a fixed-sized region. Thus the fact that a staircase configuration is not on the computational orbit of any synchronous computation doesn't mean that such a configuration can't perform an equivalent computation: the arrangement of clock spins outside of the fixed-sized region containing the circuit of interest is irrelevant as long as computation is able to proceed within this region, and as long as relative synchronization is never locally violated.

Of course what we would really like is to have a fully parallel two-dimensional system, but at least we now have shown that we can have parallelism in a computationally universal quantum Hamiltonian system with only neighbor interactions.

Conclusions

The study of the fundamental physical limits of efficient computation requires us to consider models in which the mapping between the computational and physical degrees of freedom is as close as is possible. This has led us to ask whether the structure of quantum mechanics is compatible with parallel deterministic computation. If the answer was no, then such computation would in general have to be a macroscopic phenomenon. In fact, at least in one dimension, it does seem possible to construct plausible models to simulate any locally interacting deterministic system at a constant rate and in a local manner. The problem of finding satisfactory models of fully parallel quantum computation in more than one dimension remains open.

Physically motivated models of computation such as those considered here, in which individual degrees of freedom have both a computational and a physical interpretation, act as bridges between theoretical physics and theoretical computer science. Computers constructed (for efficiency) with a physics-like structure may be usefully analyzed using concepts and techniques imported from physics (Margolus 1988); computational reinterpretations of such imported physical concepts may someday prove useful in the study of physics itself. 🖋

Acknowledgments

I would like to gratefully acknowledge conversations with R. P. Feynman in which he pointed out to me the relationship between my parallel model and his serial model, and discussions with L. M. Biafore in which it became evident that the one-dimensional version of my parallel quantum-mechanical construction might be made to run at a uniform rate.

This research was supported by the Defense Advanced Research Projects Agency and by the National Science Foundation.

REFERENCES

Benioff, P. A. 1982. "Quantum Mechanical Hamiltonian Models of Discrete Processes that Erase Their Own Histories: Application to Turing Machines." *International Journal of Theoretical Physics* 21 (3–4): 177–201. doi:10.1007/bf01857725.

———. 1986. "Quantum Mechanical Hamiltonian Models of Computers." Proceedings of the conference "New Techniques and Ideas in Quantum Measurement Theory", *Annals of the New York Academy of Sciences* 480 (1): 475–86. doi:10.1111/j.1749-6632.1986.tb12450.x.

Bennett, C. H. 1973. "Logical Reversibility of Computation." *IBM Journal of Research and Development* 17 (6): 525–32. doi:10.1147/rd.176.0525.

Deutsch, D. 1985. "Quantum Theory, the Church-Turing Principle and the Universal Quantum Computer." *Proceedings of the Royal Society A: Mathematical, Physical and Engineering Sciences* 400 (1818): 97–117. doi:10.1098/rspa.1985.0070.

Feynman, R. P. 1982. "Simulating Physics with Computers." *International Journal of Theoretical Physics* 21 (6–7): 467–88. doi:10.1007/bf02650179.

———. 1985. "Quantum Mechanical Computers." *Optics News* 11 (2): 11–20. doi:10.1364/ON.11.2.000011.

Fredkin, E., and T. Toffoli. 1982. "Conservative logic." *International Journal of Theoretical Physics* 21 (3–4): 219–53. doi:10.1007/bf01857727.

Landauer, R. 1961. "Irreversibility and Heat Generation in the Computing Process." *IBM Journal of Research and Development* 5 (3): 183–91. doi:10.1147/rd.53.0183.

Margolus, N. 1984. "Physics-Like Models of Computation." *Physica D: Nonlinear Phenomena* 10 (1–2): 81–95. doi:10.1016/0167-2789(84)90252-5.

———. 1986. "Quantum Computation." Proceedings of the conference "New Techniques and Ideas in Quantum Measurement Theory", *Annals of the New York Academy of Sciences* 480 (1): 487–97. doi:10.1111/j.1749-6632.1986.tb12451.x.

———. 1988. *Physics and Computation.* Technical report. Tech. Rep. MIT/LCS/TR-415, MIT Laboratory for Computer Science.

Peres, A. 1985. "Reversible Logic and Quantum Computers." *Physical Review A* 32 (6): 3266–76. doi:10.1103/physreva.32.3266.

Toffoli, T. 1977. "Computation and Construction Universality of Reversible Cellular Automata." *Journal of Computer and System Sciences* 15 (2): 213–31. doi:10.1016/s0022-0000(77)80007-x.

———. 1984. "Cellular Automata as an Alternative to (Rather than an Approximation of) Differential Equations in Modeling Physics." *Physica D: Nonlinear Phenomena* 10 (1–2): 117–27. doi:10.1016/0167-2789(84)90254-9.

Toffoli, T., and N. Margolus. 1987. *Cellular Automata Machines: A New Environment for Modeling.* Cambridge, MA: MIT Press.

Zurek, W. H. 1984. "Reversibility and Stability of Information Processing Systems." *Physical Review Letters* 53 (4): 391–94. doi:10.1103/physrevlett.53.391.

ملك

INFORMATION PROCESSING AT THE MOLECULAR LEVEL: POSSIBLE REALIZATIONS AND PHYSICAL CONSTRAINTS

W. G. Teich, Universität Stuttgart
and G. Mahler, Universität Stuttgart

Introduction

In the last couple of decades enormous progress has been achieved in the miniaturization of hardware elements for computing devices based on conventional microelectronics. The four-megabit chip is employed commercially, and the dimension of individual elements has already reached the submicrometer regime. But for a length scale of the order of the de Broglie wavelength (typically a few nanometers), quantization effects become important. The miniaturization process, which is limited so far by a lack of mastering of the technology to build the respective structures (mask production, etching, etc.), is bounded by fundamental physical constraints. In order to further reduce the size of the hardware elements and to increase integration, new physical concepts of information processing must be developed. Investigations in this direction can be summarized as "molecular electronics" (Carter 1982). This is concerned with information processing systems where the basic elements have a dimension of a few nanometers and, therefore, possess typical molecular properties like a discrete energy subspectrum. It is not limited to organic macromolecules as a new class of substances, but includes possible realizations in the form of semiconductor heterostructures ("quantum dots"; see Obermayer, Mahler, and Haken [1987]) as well.

Since it is difficult, if not impossible, to realize far-reaching interconnections between various subunits within a highly integrated system (Ferry and Porod 1986), molecular electronics favors a parallel architecture in the form of an array of locally interconnected cells (Teich, Obermayer, and Mahler 1988). This approach is somewhat complementary to the recent interest in novel computer architectures (hypercube architectures, neural networks, etc.), which is motivated by the limitations of the conventional von Neumann architecture. In this context cellular automata (CA) (Wolfram 1986) might be considered idealized models for a parallel computer architecture, similar to how the Turing machine is an idealized model for a sequential architecture.

In contrast to "conventional" electronic devices, which can be discussed in terms of classical transport equations, a molecular electronic system must be described by quantum mechanics ("quantum computer"). Thus, the fundamental question of computation in the regime of quantum mechanics is addressed. Contrary to some more general quantum models of computation (Deutsch 1985; Feynman 1985; Peres 1985; Zurek 1984), our starting point is realistic physical materials and interactions. In this way it is possible to discuss in detail the system and its limitations. Also, experimentally testable consequences might be found.

Physics and Function

Information processing systems can be defined by the task that they are expected to perform: a computer stores and manipulates information (Landauer 1976). To accomplish this function, the system must possess some fundamental (information-theoretic) properties. On the other hand, any computer is a physical system (Landauer 1961) and must obey the laws of physics. Therefore, we can formulate physical properties which are necessary for any physical system to be able to store and process information.

The logical basis of information processing is the *alphabet*. It consists of a finite set of distinguishable signs and is needed to represent information. The corresponding physical property is *multistability*: on the relevant time scale the system must possess several stable states (attractors). The coding of information is achieved by a one-to-one mapping of the attractors to the symbols of the alphabet. Thus, the number of attractors already limits the choice of the alphabet. In conventional microelectronic devices, multistability is achieved by various current or voltage distributions. On the molecular level, on the other hand, stationary states can be used to realize multistability. Examples are distinct structural configurations of a molecule or localized electronic states in molecules or in solid-state devices (lattice defects, impurities, quantum dots) as they are used within persistent spectral hole burning (Moerner 1988).

A computer must be able to communicate with its surroundings (input and output). Physically this corresponds to the *preparation* and *measurement* of the attractors. This implies an interaction between the system and its environment, at least during the preparation or the measurement stage. Reliable preparation and measurement is an indispensable prerequisite for using an attractor to code information. On the molecular level, this raises questions regarding quantum fluctuations and the quantum-mechanical nature of the measurement process (e.g., enhancement of quantum signals).

For the storage of information, multistability and reliable preparation and measurement of the attractors are sufficient. However, for the processing of information, further requirements can be specified. To process information, the system must divide into two or more subsystems with independent input and output (modularity) but which exchange information during the data processing. This requires a minimal amount of *control* of the system. Control means to switch selectively from one attractor to another. The necessary amount of control is given by the requirements of an independent preparation and measurement of the various subunits and the realization of simple logical operations or transition rules.

The required amount of control, and thus the modularity of the logical system, is easiest achieved by modularity of the physical system. A prerequisite for that is a hierarchy of interaction energies which allows the total quantum system to decompose into various subsystems that can be prepared and measured independently, but which still interact strongly enough to exchange the necessary amount of information. At this point the role of dissipation should be emphasized. Two quantum systems that have interacted in the past (which is necessary to process information) and evolve coherently in time cannot be separated again (Blum 1981). In order to assure the independent preparation and measurement of the subsystems, it is necessary to include dissipation, which destroys the coherence between the two subsystems.

The basis for a complex dynamics which meets these requirements is the structural complexity of the system. Neither a perfect crystal (complete order) nor an ideal gas (complete randomness) can be used to store and process information. What is necessary is a hierarchical structure of the system (Teich and Mahler 1989). The different length scales of this structure define various time and energy scales, which, on the other hand, allow the approximate decomposition into independent subsystems and thus the modular set-up of the system.

Optically Controlled Multistable Quantum System

The specific model that we consider is an optically controlled multistable quantum system, i.e., a quantum system which is coupled to a macroscopic control system (laser, photodetector) via photons. Multistability is realized in the form of localized charge-transfer excitations through a hierarchical set-up of the system. Due to localization selection rules, the charge-transfer character of the excitations inhibits the decay of the excitation and leads to vastly different time scales (Obermayer, Teich, and Mahler 1988). On the other hand, the charge transfer is used to couple different subunits (cells) via the Coulomb interaction. A specific attractor is prepared by means of a dissipative switching process, induced by a coherent light field. Selectivity of the switching processes is achieved in frequency space. Measurement of the attractors proceeds via resonance fluorescence. Control is gained by conditional switching processes in a coherent light field. Starting with the smallest functional unit, a cell, we will discuss our model in some detail in the following. A more detailed description can be found in Obermayer, Mahler, and Haken (1987), Obermayer, Teich, and Mahler (1988), Teich, Obermayer, and Mahler (1988), Teich, Anders, and Mahler (1989), and Teich and Mahler (1989).

MINIMAL MODEL FOR DISSIPATIVE SWITCHING DYNAMICS: THREE-LEVEL SYSTEM (CELL)

The minimal model for a dissipative switching dynamics, i.e., the realization of multistability (bistability in the case of the minimal model) and a reliable preparation of the attractors, is given by a three-level system (cell) with specific couplings (fig. 1): the ground state $|1\rangle$ is not (or only weakly) coupled to the first excited state $|2\rangle$ (spontaneous decay rate r_{21}), but both states $|1\rangle$ and $|2\rangle$ are strongly coupled to the excited state $|3\rangle$ (spontaneous decay rates r_{31} and r_{32}). Thus, on a time scale of $r_{21} \ll t^{-1} \ll r_{31}, r_{32}$, the excited state $|2\rangle$ is metastable, whereas the excited state $|3|\rangle$ is a transient state, since it decays into the states $|1\rangle$ and $|2\rangle$, which form the attractors of the system on this time scale. The time scale spreading $\underline{r_{21}} \ll r_{31}, r_{32}$ is the basis for the following dissipative switching dynamics: a cell is switched, e.g., from state $|1\rangle$ to $|2\rangle$ by applying a laser pulse of frequency $\omega_{31} = (E_3 - E_1)/\hbar$, where E_i is the energy of state $|i\rangle$. State $|1\rangle$ will be destabilized by this pulse, with state $|2\rangle$ remaining the only stable attractor of the system (as long as the laser pulse is on). The cell is excited to the transient state $|3\rangle$, from which it can decay spontaneously into state $|2\rangle$ or back into state $|1\rangle$ (where it will be re-excited again as long as the laser pulse is on).

Since the switching dynamics is a stochastic process, it must be described in terms of a transition probability (Teich and Mahler 1989), which is determined by the intensity and duration of the applied laser pulse. Selectivity in real space cannot be achieved (the extension of a cell is much smaller than the wavelength of the applied laser pulse), so that the transition from state $|2\rangle$ to state $|3\rangle$ is always driven as well, albeit being off-resonant (back-reaction). Therefore, for long pulse durations the transition probability approaches an asymptotic value, which depends on the intensity of the laser pulse and the selectivity in frequency space, i.e., on the difference between the two transition frequencies ω_{31} and ω_{32}. The larger this difference is, the closer the asymptotic transition

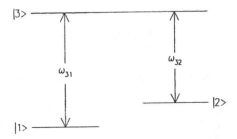

Figure 1. Quantum optical model for a single cell A. The coupling between the states is indicated by the "overlap" of the respective levels (see text).

probability approaches unity. In realistic systems the average error probability, i.e., the difference between the transition probability and 1, might be as small as 10^{-10} (Teich and Mahler 1989). The dissipative character of the switching dynamics assures that the switching process is not sensitive to initial conditions and incorrect light pulse parameters (frequency, intensity, pulse length).

For special initial conditions (i.e., starting from "pure* attractors $|1\rangle$ or $|2\rangle$ and not from a coherent superposition of both) and when only one of the two resonance frequencies ω_{31} and ω_{32} is applied at a time, the dynamics of the cell can be characterized by a machine table (table 1). It gives initial and final states and the frequency required to accomplish the desired switching process.

Each excitation of the cell is accompanied by a charge transfer, i.e., each of the three states $|1\rangle$, $|2\rangle$, and $|3\rangle$ has a

	$|1\rangle$	$|2\rangle$
$|1\rangle$	0	ω_{31}
$|2\rangle$	ω_{32}	0

Table 1. Machine table for a single cell A.

different static dipole moment. This will be important for the coupling of two cells and for the formation of a network of coupled cells.

The measurement process is realized via resonance fluorescence. A fourth state $|4\rangle$, which is strongly coupled to the ground state $|1\rangle$ but does not couple to the metastable state $|2\rangle$ ($r_{42} \ll r_{41}$), is added to a cell (Teich, Anders, and Mahler 1989). Applying a laser pulse of frequency ω_{41}, the cell will only scatter photons of the same frequency if it is in state $|1\rangle$. Being in state $|2\rangle$, no photons will be detected. Again, the basis for the amplification of the quantum signal is the time scale spreading, $r_{42}, r_{21} \ll r_{41}$, which allows one to perform a repeated scattering experiment during the lifetime of the metastable state $|2\rangle$, and in this way one arrives at a reliable predicate about the quantum state of the system.

Possible realizations of such a three- or four-level system include three-dimensional semiconductor heterostructures (double-quantum-dot model—see Obermayer, Mahler, and Haken [1987] and Obermayer, Teich, and Mahler [1988]) and complex macromolecules like donator–acceptor molecules. In either case, the basis for the relevant subspectrum with the appropriate couplings is a hierarchical structure of the system (Teich and Mahler 1989) which allows us to tailor selection rules via localization of the wave functions (Obermayer, Teich, and Mahler 1988). An even smaller realization of such a three-level system is given by a single trapped ion (Nagourney, Sandberg, and Dehmelt 1986). In this case the corresponding coupling between the states is due to symmetry-based selection rules. However, the various states of a single trapped ion do not possess different dipole moments and, therefore, cannot be coupled via the dipole–dipole interaction as described below.

Applications include an array of uncoupled cells which

could be used as an information storage device, similar to persistent hole burning (Moerner 1988). Storage capacities of up to 10^9 bits/cm^2 might be achieved in this way (Obermayer, Teich, and Mahler 1988).

MINIMAL MODEL FOR CONDITIONAL SWITCHING DYNAMICS: TWO COUPLED CELLS

Control of a quantum system can be demonstrated with the minimal model for a conditional switching dynamics. We ~427~ consider two cells A and B (fig. 2) with four distinct transition frequencies ω_{31}, ω_{32}, ω_{64}, and ω_{65}, i.e., each transition can be addressed selectively in frequency space (due to the small dimensions of the system, selectivity in real space is not possible). The two cells are separated by a distance R, which is supposed to be large enough to assure localized excitations within each cell. Multistability is then a fixed property of each cell. On the other hand, R must be small enough in order for the two cells to interact via the Coulomb interaction. The Coulomb interaction between the charge distributions of the two cells leads to an energy renormalization which, to lowest order, is given by the dipole–dipole interaction. Since the ground state and excited states of each cell have different dipole moments, the transition frequencies of each cell depend on the state of the other cell. The splitting of the transition frequencies is given by (Teich, Obermayer, and Mahler 1988)

$$\hbar\Delta\omega = \hbar\omega_{31}(4) - \hbar\omega_{31}(5) = \frac{p_A p_B}{4\pi\epsilon\epsilon_0 R^3} F(\Theta_A, \Theta_B) \quad (1)$$

and can be as large as $1\,\mathrm{meV}$ for a semiconductor hetero-structure. $\omega_{31}(i)$ is the conditional transition frequency between states $|1\rangle$ and $|3\rangle$ of cell A, if cell B is in state $|i\rangle$. p_A and p_B measure the magnitudes of the change of the static dipole moment between states $|1\rangle$ and $|2\rangle$ (cell A) or $|4\rangle$ and $|5\rangle$ (cell B), and the angular factor $F(\Theta_A, \Theta_B)$ results from the

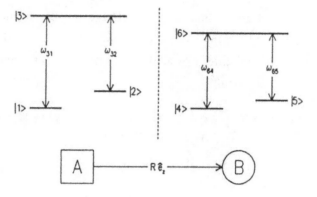

Figure 2. Schematic drawing of two coupled cells A and B with different transition frequencies, separated by a distance R.

direction of the charge transfer in cells A and B: $F(\Theta_A, \Theta_B) = 1$ for a charge transfer parallel to the direction between the two cells, i.e., the z-direction (see fig. 2). $\epsilon\epsilon_0$ is the static dielectric constant of the embedding material.

A conditional switching dynamics can be induced, e.g., in cell A, by applying a laser pulse of frequency $\omega_{31}(4)$ and small enough bandwidth: cell A is switched from state $|1\rangle$ to state $|2\rangle$ if cell B is in state $|4\rangle$ and remains stationary if cell B is in state $|5\rangle$. The system can be switched selectively from state $|1\rangle|4\rangle$ to state $|2\rangle|4\rangle$. All other states remain practically unaffected for large enough frequency selectivity $\Delta\omega$. Similar to the case of a single cell, a reduced description of the dynamics in the form of a conditional machine table for each cell can be given (table 2). Again, the machine tables are only valid for purely local excitations of the system. Also, applying frequencies which can simultaneously lead to transitions in cell A and cell B is not allowed.

The machine table (table 2) describes the possible control over the system. Using two simultaneous laser pulses with frequencies $\omega_{31}(4)$ and $\omega_{31}(5)$, it is possible to prepare state

	cell A			cell B	
B	$\lvert 1\rangle \to \lvert 2\rangle$	$\lvert 2\rangle \to \lvert 1\rangle$	A	$\lvert 4\rangle \to \lvert 5\rangle$	$\lvert 5\rangle \to \lvert 4\rangle$
$\lvert 4\rangle$	$\omega_{31}(4)$	$\omega_{32}(4)$	$\lvert 1\rangle$	$\omega_{64}(1)$	$\omega_{65}(1)$
$\lvert 5\rangle$	$\omega_{31}(5)$	$\omega_{32}(5)$	$\lvert 2\rangle$	$\omega_{64}(2)$	$\omega_{65}(2)$

Table 2. Conditional machine table for two coupled cells A and B.

$\lvert 2\rangle$ in cell A independently of the state of cell B. Similarly, all other states can be prepared independently. By applying a laser pulse with the single frequency $\omega_{32}(4)$, the new state of cell A depends on the old states of cell A and cell B. For suitable coding this mapping represents a logical "OR." Similarly, all other elementary logical functions can be realized (Teich, Obermayer, and Mahler 1988).

MINIMAL MODEL FOR AN ADAPTIVE SYSTEM: ONE-DIMENSIONAL CELLULAR STRUCTURE

As an example of a network of coupled cells, we consider a linear arrangement of alternating cells A and B (fig. 3). The A–B repeat unit is necessary in order to achieve conditional dynamics with (left and right) nearest-neighbor coupling. If a cell, e.g., A, is to be switched depending on the state of neighboring cells, these neighboring cells must be passive during the finite switching time and must, therefore, be physically distinct from cell A, at least regarding the transition frequencies.

As for the case of two coupled cells, the conditional dynamics is realized via state-dependent transition frequencies

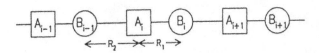

Figure 3. Real-space model for a linear chain of alternating cells A and B. The distances to the left and right neighbors are in general different ($R_1 \neq R_2$).

due to the dipole–dipole interaction (see fig. 4). Since only nearest-neighbor coupling is desired, the influence of all other cells has to be compensated for. This can be achieved by a large enough bandwidth $\delta\omega$ of the laser pulse, which, however, must be smaller than the separation $\Delta\omega$ between different frequency bands. In this case the transition probability for each cell depends only on the state of the adjacent cells. The required bandwidth can be found to be (Teich, Obermayer, and Mahler 1988)

$$\delta\omega \approx 0.6\,\Delta\omega. \tag{2}$$

In order to achieve distinct transition frequencies for each of the four possible configurations of nearest neighbors, the frequency shifts of the left and right neighbors must be different. Due to the R^{-3} dependence of the dipole–dipole interaction, this can be realized by an asymmetric arrangement of the cells, i.e., the distances to the left and the right neighbors are different. This symmetry-breaking physically defines a direction on the chain which, on the other hand, is necessary to get a directed information flow in the system.

Since individual cells cannot be addressed selectively either in real space or in frequency space, the preparation of the cellular structure must be performed with the help of a shift operation. Starting from a physical inhomogeneity (i.e., a cell D with transition frequencies distinct from those of cells A and B), any inhomogeneous state can be prepared: a temporal pattern (successive preparations of cell D) is transformed by successive shift operations into a spatial pattern of the cellular structure (serial input; see Teich, Obermayer, and Mahler [1988]). Similarly, the state of the one-dimensional cellular structure can be measured.

If the angular dependence of the dipole–dipole interaction is exploited (i.e., the direction of the charge transfer changes

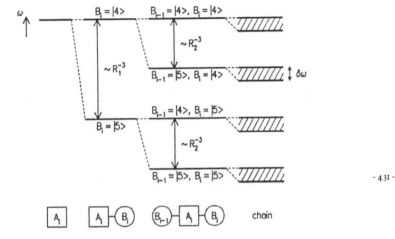

Figure 4. Dependence of the transition frequency on the state of neighboring cells due to the dipole–dipole interaction (see text). The hatched area indicates the influence of all but the adjacent cells.

from one cell to the other), the ratio between the bandwidth $\delta\omega$ and the frequency selectivity $\Delta\omega$ can be reduced further. In this case a coupling between next nearest neighbors can also be achieved. The transition frequencies depend on the state of the nearest two left and right neighbors, and the unit cell has to be extended to three cell types A, B, and C.

For special initial conditions (direct product of localized states) and for a specific choice of the laser pulse sequences (i.e., cells A and B are not switched simultaneously), the dynamics of the linear chain can be characterized by a local transition table for each cell type A and B (table 3). Similarly to CA (Wolfram 1986), the global evolution of the cellular structure is fixed in this way, and the final state might not be deduced but by direct simulation of the dynamical evolution.

The dynamical evolution of the system adapts to various stimulations by its environment (i.e., laser pulse sequences). In this way the behavior of the one-dimensional cellular structure varies from the deterministic dynamics of a one-dimensional

		cell A_i	
B_{i-1}	B_i	$\lvert 1\rangle \to \lvert 2\rangle$	$\lvert 2\rangle \to \lvert 1\rangle$
$\lvert 4\rangle$	$\lvert 4\rangle$	$\omega_{31}(4,4)$	$\omega_{32}(4,4)$
$\lvert 4\rangle$	$\lvert 5\rangle$	$\omega_{31}(4,5)$	$\omega_{32}(4,5)$
$\lvert 5\rangle$	$\lvert 4\rangle$	$\omega_{31}(5,4)$	$\omega_{32}(5,4)$
$\lvert 5\rangle$	$\lvert 5\rangle$	$\omega_{31}(5,5)$	$\omega_{32}(5,5)$
		cell B_i	
A_i	A_{i+1}	$\lvert 4\rangle \to \lvert 5\rangle$	$\lvert 5\rangle \to \lvert 4\rangle$
$\lvert 1\rangle$	$\lvert 1\rangle$	$\omega_{64}(1,1)$	$\omega_{65}(1,1)$
$\lvert 1\rangle$	$\lvert 2\rangle$	$\omega_{64}(1,2)$	$\omega_{65}(1,2)$
$\lvert 2\rangle$	$\lvert 1\rangle$	$\omega_{64}(2,1)$	$\omega_{65}(2,1)$
$\lvert 2\rangle$	$\lvert 2\rangle$	$\omega_{64}(2,2)$	$\omega_{65}(2,2)$

Table 3. Local transition rules for cell types A and B.

unidirectional CA (Teich, Obermayer, and Mahler 1988) to the stochastic dynamics of, e.g., the one-dimensional kinetic Ising model (Teich and Mahler 1989).

Summary

For the example of an optically controlled multistable quantum system, we have demonstrated the connection

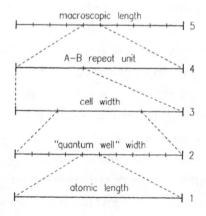

Figure 5. Five hierarchical levels, defined by various length scales of the system.

between a complex hierarchical structure and the complex dynamics of the system. Different length scales define various hierarchical levels of the system (fig. 5). The number of hierarchical levels of the system (in our case five) is a measure of the "homogeneous" complexity of the system. The minimum number of five hierarchical levels is a prerequisite in order to realize multistability, preparation, measurement, and control, necessary for achieving a complex dynamics which is equivalent to information processing. Neither a perfect crystal nor an ideal gas, which both possess only two hierarchical levels (a macroscopic length scale and an atomic length scale), fulfill these requirements. ✒

~ 433 ~

Acknowledgment

Financial support by the Deutsche Forschungsgemeinschaft (Sonderforschungsbereich 329) is gratefully acknowledged.

REFERENCES

Blum, K. 1981. "Coupled Systems." In *Density Matrix Theory and Applications,* 63–84. New York, NY: Plenum Press. doi:10.1007/978-1-4615-6808-7_3.

Carter, F. L., ed. 1982. *Molecular Electronic Devices.* New York, NY: Marcel Dekker.

Deutsch, D. 1985. "Quantum Theory, the Church-Turing Principle and the Universal Quantum Computer." *Proceedings of the Royal Society A: Mathematical, Physical and Engineering Sciences* 400 (1818): 97–117. doi:10.1098/rspa.1985.0070.

Ferry, D. K., and W. Porod. 1986. "Interconnections and Architecture for Ensembles of Microstructures." *Superlattices and Microstructures* 2 (1): 41–44. doi:10.1016/0749-6036(86)90151-5.

Feynman, R. P. 1985. "Quantum Mechanical Computers." *Optics News* 11 (2): 11–20. doi:10.1364/ON.11.2.000011.

Landauer, R. 1961. "Irreversibility and Heat Generation in the Computing Process." *IBM Journal of Research and Development* 5 (3): 183–91. doi:10.1147/rd.53.0183.

———. 1976. "Fundamental Limitations in the Computational Process." *Berichte der Bunsengesellschaft für physikalische Chemie* 80 (11): 1048–59. doi:10 . 1002 / bbpc . 19760801105.

Moerner, W. E. 1988. *Persistent Spectral Hole-Burning: Science and Applications*. Berlin, Germany: Springer-Verlag. doi:10.1007/978-3-642-83290-1.

Nagourney, W., J. Sandberg, and H. Dehmelt. 1986. "Shelved Optical Electron Amplifier: Observation of Quantum Jumps." *Physical Review Letters* 56 (26): 2797–99. doi:10.1103/physrevlett.56.2797.

Obermayer, K., G. Mahler, and H. Haken. 1987. "Multistable Quantum Systems: Information Processing at Microscopic Levels." *Physical Review Letters* 58 (20): 1792–95. doi:10.1103/physrevlett.58.1792.

Obermayer, K., W. G. Teich, and G. Mahler. 1988. "Structural Basis of Multistationary Quantum Systems. I. Effective Single-Particle Dynamics." *Physical Review B* 37 (14): 8096–110. doi:10.1103/physrevb.37.8096.

Peres, A. 1985. "Reversible Logic and Quantum Computers." *Physical Review A* 32 (6): 3266–76. doi:10.1103/physreva.32.3266.

Teich, W. G., G. Anders, and G. Mahler. 1989. "Transition between Incompatible Properties: A Dynamical Model for Quantum Measurement." *Physical Review Letters* 62 (1): 1–4. doi:10.1103/physrevlett.62.1.

Teich, W. G., and G. Mahler. 1989. "Optically Controlled Multistability in Nanostructured Semiconductors." *Physica Scripta* 40 (5): 688–93. doi:10.1088/0031-8949/40/5/019.

Teich, W. G., K. Obermayer, and G. Mahler. 1988. "Structural Basis of Multistationary Quantum Systems. II. Effective Few-Particle Dynamics." *Physical Review B* 37 (14): 8111–21. doi:10.1103/physrevb.37.8111.

Wolfram, S. 1986. *Theory and Applications of Cellular Automata*. Singapore: World Scientific.

Zurek, W. H. 1984. "Reversibility and Stability of Information Processing Systems." *Physical Review Letters* 53 (4): 391–94. doi:10.1103/physrevlett.53.391.

ơjꟼ

HOW CHEAP CAN MECHANICS'
FIRST PRINCIPLES BE?

Tommaso Toffoli, Massachusetts Institute of Technology

It is fashionable today to explain various phenomenological laws as emergent properties of appropriate collective behavior. Here, we argue that the very fundamental principles of physics show symptoms of being emergent properties, and thus beg for further reduction.

1. Introduction

One often speaks, generically, of "the laws of physics." The physicist, however, is well aware that different kinds of laws each have a different status, and according to their status are meant to play a different role in both theory and applications. The major status categories are roughly as follows:

- *Analytical mechanics.* Here we have the "constitution" of physics—the principles of classical mechanics, relativity, and quantum mechanics. When we say, "Let's consider a Hamiltonian of this form," we do not pretend that a physical system governed by such a law actually exists; we merely imply that the law would not be struck down by physics' supreme court as "unconstitutional."

- *Fundamental processes.* Here we have those physical interactions that are actually observed and that presumably belong to nature's most fundamental repertoire. They are the "op-codes" (using Margolus's metaphor; see Margolus [1988]) which the supreme architect actually decided to include in physics' "machine language." We tentatively assume that, as in the design of a computer chip, other choices of op-codes were potentially available and could have been equally effective.

Of course, some "grand unified theory" may later show that what appeared to be independent choices at the op-code level are actually forced consequences of a single master choice. Moreover, we may realize that what we thought was a primitive op-code is actually implemented as a higher-level construct—a "subroutine call." But we are all familiar with this kind of issue from experience with man-made worlds.[1]

• *Statistical mechanics.* Here we have laws that emerge out of the collective behavior of a large number of elements. The quantities involved in these laws may not even be meaningful for individual systems or experiments. Intuitively, one may expect that almost every detail of the microscopic interactions will be washed out by macroscopic averaging; only features that are supported by a definite conspiracy (such as a particular symmetry or conservation law) will bubble up all the way to the macroscopic surface and emerge as recognizable statistical laws.

In the past few decades, an enormous range of complex physical phenomena have been successfully explained as inexorable statistical-mechanical consequences of known fundamental processes or plausible stylizations thereof. Without doubt, the

---

[1] The choice of op-codes for, say, the IBM/360 family of computers reveals strong constraints of economy, consistency, and completeness. And in the cheapest models of this family many of the op-codes documented in the machine-language manual are emulated by software traps rather than directly implemented in hardware; the timing may be different, but the logic is identical.

reduction of phenomenology to fundamental processes via statistical mechanics is today one of the most productive paradigms (cf. Kuhn 1970) of mathematical physics. Explaining the texture of mayonnaise has become a likely subject for ten articles in *Physical Review*, and no one would be surprised if its mathematics turned out to be isomorphic to that needed to explain the fine structure of quarks.

This work on collective phenomena has revealed principles that appear to have a universal and fundamental character not unlike that of the principles of mechanics. In this paper, we shall turn the tables and ask, "Are perhaps the very *principles of mechanics* so universal and fundamental just because they are emergent aspects of an extremely fine-grained underlying structure, and thus chiefly *mathematical* rather than *physical* in content?"

A coin, no matter what its composition, shape, or tossing technique, can be characterized by a single real parameter k such that over a large number of trials it will come up heads close to a fraction k of the time. The existence of such a parameter is not usually regarded as a property of our physical world *per se*—a choice made by God when establishing the laws of nature; rather, it is seen as a mathematical consequence of almost any choice about physics God could have made.

In the same vein, one would like to ask whether, for instance, the laws of mechanics are symplectic because God explicitly decided to make them so, or whether this symplectic character automatically follows from virtually any reasonable choice of fine-grained first principles. Similarly, can one think of simple ground rules for physics whereby relativity would appear as little surprising as the law of large numbers?

In this paper, we shall give some circumstantial evidence that questions of the above kind are scientifically legitimate and intellectually rewarding. Namely, we'll look at a number of physical concepts that are usually regarded as primitive, and in each case we'll show a plausible route for reduction to much simpler concepts.

2. Continuity

In both the classical and the quantum description, the state of a physical system evolves as a *continuous* function of time. In mathematics, it is well known that certain discrete constructs (e.g., the distribution of prime numbers, $\pi(n)$) can be approximated by continuous ones (in this example, the Riemann function $R(x)$). However, continuity does not *invariably* emerge from discreteness through some universal and well-understood mechanism, so that, when it does, we are justified in asking *why*. Once we understand the reasons in one case, we may hope to derive sufficient conditions for its emergence in a more general situation. Here we'll give an example of sufficient conditions in a kinematical context.

Consider an indefinitely extended two-dimensional lattice of spacing λ, having a 1 ("particle") or a 0 ("vacuum") at each site, as in figure 1(a). As we move, say, along the x axis, the microscopic *density function* $\rho(x, y)$ will display the discontinuous behavior of figure 1(b).

Let us define a whole sequence ρ_n of new density functions, with $\rho_n(x, y)$ denoting the average density over the square window of side $n\lambda$ centered at (x, y). For example, ρ_3 can take any of the ten values $0, 1/9, 2/9, \ldots, 8/9, 1$ (fig. 2(a)). However, as x increments by λ—and the corresponding window slides one lattice position to the right—ρ_3 cannot arbitrarily jump between any two of these values; the maximum size of a jump is $1/3$ (fig. 2(b)).

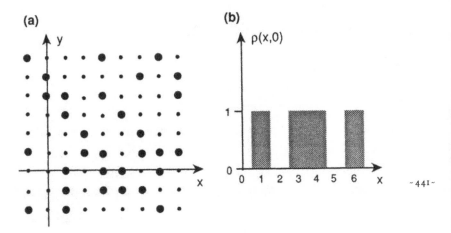

Figure 1. (a) Particles on a lattice. (b) Density plot along a line $y = $ constant.

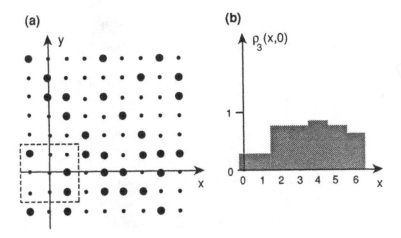

Figure 2. (a) A 3×3 window centered on (x, y). (b) Average-density plot, as the center of the window moves along a line $y = $ constant.

In general, while ρ_n depends on the number of particles contained in the entire window (*volume* effect), the change $\Delta\rho_n$ corresponding to $\Delta x = \pm\lambda$ depends only on the number of particles swept by the edge of the window (*surface* effect); thus,

$$|\Delta\rho_n| \leq \frac{n}{n^2} = \frac{1}{n}, \tag{1}$$

and

$$\lim_{n\to\infty} \Delta\rho_n = 0. \tag{2}$$

If now we let the lattice spacing λ decrease in the same proportion as n increases (so that the area of the window remains constant), in the limit as $n \to \infty$ the sequence ρ_n converges to a *uniformly continuous* function of x and y.

The above considerations involving a static configuration of particles are trivial. Now, let us introduce an arbitrary discrete dynamics (τ will denote the time spacing between consecutive states), subject only to the following constraints:

- *Locality.* The state of a site at time $t + \tau$ depends only on the state at time t of the neighboring sites.

- *Particle conservation.* The total number of particles is strictly conserved.

In one time step, only particles that are lying next to the window's border can move in or out of the window: much as that on x, the dependency of ρ_n on t is also a surface effect. If in taking the above limit we let the time spacing τ shrink in the same proportion as the lattice spacing λ, so as to leave the "speed of light" (one site per step) constant, the sequence $\rho_n(x, y; t)$ converges to a uniformly continuous function of t.

Note that if either locality or particle conservation did not hold, ρ_n as a function of *time* would not, in general, converge to a definite limit. Thus, we have characterized a situation where the emergence of a continuous dynamics is reducible to certain general properties of a (conceptually much simpler) underlying fine-grained dynamics.

Is that the store where physics buys continuity? Who knows—but Occam would say it's a good bet!

3. Variational Principles

In order to explicitly construct the evolution of an arbitrary dynamical system over an indefinitely long stretch of time, one needs laws in *vectorial* form. In the discrete-time case, a vectorial law gives the *next state* u_{t+1} of the system as a function of the *current state* u_t,

$$u_{t+1} = F u_t ; \qquad (3)$$

though in many cases of interest F can be captured by a more   -443- concise algorithm, full generality demands that F be given as an exhaustive lookup table, since its values for different values of u can in principle be completely arbitrary.

In the continuous case, a vectorial law gives the *rate of change* of the current state $u(t)$,

$$\frac{d}{dt} u = f u(t), \qquad (4)$$

where f can be thought of as a lookup table having a continuum of entries rather than a discrete set of entries.

Vectorial laws of the form in equation 3 or 4 are very general, and can be used to describe systems that have nothing to do with physics. Only a subset of such laws, characterized by particular constraints on the form of F or f, will describe admissible physical systems. Thus, as long as we restrict our attention to physical systems, the lookup tables used in equations 3 and 4 have less than maximal algorithmic entropy, and can in principle be compressed. For example, in a Hamiltonian system with one degree of freedom, the state can be written as an ordered pair $u = \langle q, p \rangle$ in such a way that, instead of two lookup tables f and g as in the hypothetical general case

$$\begin{cases} \dfrac{d}{dt} q = f \langle q, p \rangle, \\[2mm] \dfrac{d}{dt} p = g \langle q, p \rangle, \end{cases} \qquad (5)$$

one only needs a *single* lookup table H, as in the well-known Hamilton equations

$$\begin{cases} \dfrac{d}{dt}q = \dfrac{d}{dp}H\langle q,p\rangle, \\[2mm] \dfrac{d}{dt}p = -\dfrac{d}{dq}H\langle q,p\rangle. \end{cases} \qquad (6)$$

This compression factor of 2 (or of $2n$ for n degrees of freedom) is attained at a cost. To obtain the current value of dq/dt, it is no longer enough to look at a single entry of a table, as in equation 5; in fact, one has to determine the trend of H for variations of p in the vicinity of the current value of (q,p), and this entails looking up a whole range of entries. Compressed data save memory space, it is true, but entail more computational work.

3.1 $T = dS/dE$ FOR ALMOST ALL SYSTEMS

In general, there appear to be strong constraints on the form of admissible physical laws; these constraints are often best expressed by variational principles. Could it be that the actual constraints are much weaker, and that the stronger constraints that we see are the result of our way of *perceiving* these laws— perhaps through heavy statistical averaging?

One of the simplest variational principles of physics is the relation $T = dS/dE$, where T denotes period, E energy, and S action. Here we show that for the *most general* class of discrete, invertible dynamical systems, the typical class element still obeys the same relation; yet in these systems the dynamics is defined by an *arbitrary* permutation of the state set! One may wonder, then, whether this relation appears in physics as a consequence of combinatorial principles of a general nature— much as the law of large numbers—rather than as the expression of physics-specific principles.

In Newtonian mechanics, consider a conservative system with one degree of freedom. Let T be the period of a given orbit

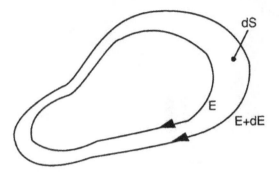

Figure 3. An energy variation, dE, and the corresponding action variation, dS.

of energy E, and let dS be the volume of phase space swept when the energy of the orbit is varied by an amount dE (fig. 3). As is well known (see, e.g., Arnold 1978), these quantities obey the relation

$$T = \frac{dS}{dE}. \qquad (7)$$

Quantities analogous to T, dE, and dS can be defined for dynamical systems of a much more general nature. Under what conditions will the above relation still hold? We shall show that equation 7 is a statistical consequence of very weak assumptions on the structure of the dynamics.

We shall consider the class χ_N consisting of *all* discrete systems having a finite number N of states and an invertible but otherwise arbitrary dynamics. We may assume that continuous quantities arise from discrete ones in the limit $N \to \infty$. In general, equation 7 will not hold for each individual system; however, if one looks at the class as a whole, one may ask whether this relation holds approximately for most systems of the class. Alternatively, one may ask whether this relation holds for a suitably defined "average" system—treated as a representative of the whole class.

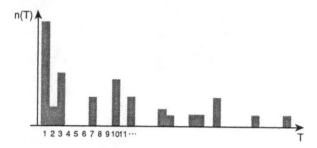

Figure 4. Orbit-length histogram.

A similar approach is widely used in statistical mechanics.[2] In our context, though, statistical methods are applied to "ensembles" in which the missing information that characterizes the ensemble concerns a system's *law* rather than its initial *state*.

The ensemble χ_N consists of $N!$ systems—i.e., all possible permutations of N states. Systems of this kind have very little structure; nonetheless, one can still recognize in them the "precursors" of a few fundamental physical quantities. For instance, the period T of an orbit is naturally identified with the number of states that make up the orbit. Likewise, a volume S of state space will be measured in terms of how many states it contains. It is a little harder to identify a meaningful generalization of energy; the arguments presented in the next subsection suggest that in this case the correct identification is $E = \log T$, and this is the definition that we shall use below.

Armed with the above "correspondence rules," we shall investigate the validity of equation 7 for the ensemble χ_N.

Each system of χ_N will display a certain distribution of orbit lengths; that is, one can draw a histogram showing, for

---

[2]For example, in an ideal gas, almost all the systems in an ensemble at a given temperature display a velocity distribution that is very close to the Boltzmann distribution; the latter can thus be taken as the "representative" distribution for the whole ensemble, even though hardly any element of the ensemble need follow that distribution *exactly*.

$T = 1, \ldots, N$, the number $n(T)$ of orbits of length T (see fig. 4). If in this histogram we move from abscissa T to $T + dT$ we will accumulate a count of $n(T)\,dT$ orbits. Since each orbit contains T points, we will sweep an amount of state space equal to $dS = T\,n(T)\,dT$; thus

$$\frac{dS}{dT} = T\,n(T). \tag{8}$$

On the other hand, since $E = \log T$, we have

$$\frac{dT}{dE} = T; \tag{9}$$

hence

$$\frac{dS}{dE} = \frac{dS}{dT}\frac{dT}{dE} = T^2 n(T). \tag{10}$$

Therefore, the original relation in equation 7 will hold if and only if the orbit-length distribution is of the form

$$n(T) = \frac{1}{T}. \tag{11}$$

Do the systems of χ_N display this distribution?

Observe that as N grows, the number of systems in χ_N grows much faster than the number of possible orbit-length distributions: most distributions will occur many times, and certain distributions may appear with a much greater frequency than others. Indeed, as $N \to \infty$, almost all of the ensemble's elements will display a similar distribution. In such circumstances, the "typical" distribution is just the mean distribution over the ensemble, denoted by $\overline{n(T)}$.

It turns out that for χ_N the mean distribution is exactly

$$\overline{n_N(T)} = \frac{1}{T} \tag{12}$$

for any N.

In fact, we construct a specific orbit of length T by choosing T states out of N and arranging them in a definite circular sequence. This can be done in $\binom{N}{T}T!/T$ different ways. To know in how many elements of the ensemble the orbit thus constructed occurs, we observe that the remaining $N - T$ elements can be connected in $(N - T)!$ ways. Thus, the total number of orbits of length T found anywhere in the ensemble is

$$\binom{N}{T}\frac{T!}{T} \cdot (N - T)! = N!\frac{1}{T}. \tag{13}$$

Divide by the size $N!$ of the ensemble to obtain $1/T$.

Thus, the typical system of χ_N obeys equation 7. Intuitively, when N is large enough to make a continuous treatment meaningful, the odds that a system picked at random will appear to be governed by the variational principle $T = dS/dE$ are overwhelming.

3.2 WHY $E = \log T$

Finite systems lack the rich topological structure of the state space found in analytical mechanics. Beside invertibility, in general the only *intrinsic*[3] structure that they are left with is the following:

> Given two points a and b, one can tell whether b can be reached from a in t steps; in particular (for $t = 0$), one can tell whether or not $a = b$.

Thus, for instance, one can tell how many orbits of period T are present, but of these one cannot single out an individual one without actually pointing at it, because they all "look the same."

To see whether there is a quantity that can be meaningfully called "energy" in this context, let us observe that physical

---

[3]That is, independent of the labeling of the points and thus preserved by any isomorphism.

energy is a function E, defined on the state space, having the following fundamental properties:

1. *Conservation.* E is constant on each orbit (though it may have the same value on different orbits).

2. *Additivity.* The energy of a collection of weakly coupled system components equals the sum of the energies of the individual components.

-449-

3. *Generator of the dynamics.* Given the constraints that characterize a particular class of dynamical systems, knowledge of the function E allows one to uniquely reconstruct, *up to an isomorphism*, the dynamics of an individual system of that class.

The proposed identification $E = \log T$ obviously satisfies property 1.

As for property 2, consider a finite system consisting of two independent components, and let a_0 and a_1 be the respective states of these two components. Suppose for a moment that a_0 is on an orbit of period 3 and a_1 on one of period 7; then the overall system state $\langle a_0, a_1 \rangle$ is on an orbit of length 21, i.e., $\log T = \log T_0 + \log T_1$. This argument would fail if T_0 and T_1 were not coprime. However, for randomly chosen integers the expected number of common factors grows extremely slowly with the size of the integers themselves (Schroeder 1986), so that approximate additivity holds almost always.

As for property 3, an individual system of χ_N is completely identified—up to an isomorphism—by its distribution $n(T)$, and thus any "into" function of T (in particular, $E = \log T$) satisfies this property.

Note that the ensemble χ_N consists of *all* invertible systems on a state space of size N. If we placed further constraints on the make-up of the ensemble, i.e., if we restricted our attention to a subset of systems having additional structure, some of the above arguments may cease to be valid. For example, while it is true that for large N *almost all* subensembles of χ_N retain distribution (eq. 12), in a few "perverse" cases the distribution will substantially depart from $1/T$ and, if we still assume that $E = \log T$, equation 7 may fail to hold. Moreover, systems that were isomorphic within χ_N may no longer be so when more structure is introduced; to allow us to tell that two systems are intrinsically different, the energy function may have to be "taught" to make finer distinctions between states than just on the basis of orbit length. But all this is besides the point we are making here; a fuller discussion of these issues can be found in Toffoli (1989a).

3.3 CONCLUSIONS

The fact that a specific variational principle of mechanics emerges quite naturally, via statistical averaging, from very weak information-mechanical assumptions does not tell us much about what fine-grained structure, if any, may actually underlie traditional physics; the relevant point is that we come to recognize that such a principle happens to be of the right form to be an emergent feature. When we see a Gaussian distribution in a sequence of heads and tails, we can't really tell *what* coin is being tossed, but conceptual economy will make us guess that somebody is tossing *some kind* of coin, rather than concocting the sequence by explicit use of the Gaussian function.

4. Relativity

The fact that the physics of flat spacetime is Lorentz, rather than Galilean, invariant is usually treated as an independent postulate of physics, much as Euclid's fifth axiom in geometry. In other words, God could have chosen differently; Lorentz invariance has to be acknowledged, not derived.

However, if we look at the most naive models of distributed computation, we see that Lorentz invariance naturally emerges as a *statistical* feature and admits a very intuitive information-mechanical interpretation. Much as in the previous section, we do not want to claim that this is the way relativity comes about in nature; we just want to stress that the mathematics of relativity happens to lie in one of those universality classes that arise from collective phenomena.

4.1 ORIENTATION

Consider the two-dimensional random walk on the (x, y) lattice. At the microscopic level, this dynamics is not rotation invariant (except for multiples of a quarter-turn rotation); however, invariance under the continuous group of rotations *emerges* at the macroscopic level (fig. 5). In fact, for $r^2 = x^2 + y^2 < t$ and in the limit as $t \to \infty$, the probability distribution $P(x, y; t)$ for a particle started at the origin converges to

$$\frac{1}{2\pi t} e^{-\frac{1}{2} \frac{x^2 + y^2}{t}}, \qquad (14)$$

i.e., depends on x and y only through $x^2 + y^2 = r^2$.

Now, there is a strict formal analogy between a *circular rotation* by an angle ϕ in the (x, y) plane and a Lorentz transformation with velocity β in the (t, x) plane—which can be written as a *hyperbolic rotation* by a rapidity $\theta = \tanh^{-1} \beta$:

$$\begin{bmatrix} t \\ x \end{bmatrix} = \begin{bmatrix} \cosh \theta & \sinh \theta \\ \sinh \theta & \cosh \theta \end{bmatrix} \begin{bmatrix} t' \\ x' \end{bmatrix}. \qquad (15)$$

Riding on this analogy, one may hope to find a microscopic dynamics on the (t, x) lattice for which Lorentz invariance

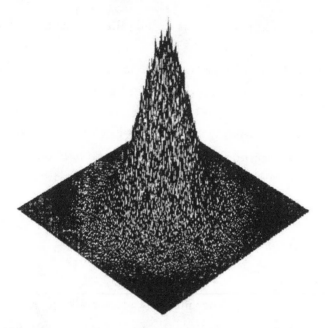

Figure 5. In the two-dimensional random walk on a lattice, circular symmetry naturally emerges at the macroscopic level out of a square-symmetry microscopic law.

(which is out of the question at the microscopic level) would emerge at the macroscopic level.

Let's look first at the one-dimensional random walk on a lattice, with probability p of moving to the right and probability $q = 1 - p$ of moving to the left. For $p = q = 1/2$, the evolution of the resulting binomial distribution is characterized, macroscopically, by a mean $\mu = 0$ and a standard deviation $\sigma = \sqrt{t/4}$ (fig. 6(a)).

In general, $\mu = (p-q)t$. If we shift the parameter p away from its center value of $1/2$, the center of mass of the distribution will start moving at a uniform velocity $\beta = p - q$. Let's try to offset this motion by a Galilean transformation

$$x = x' + \beta t'. \tag{16}$$

Macroscopically, the new system will evolve, in the new frame,

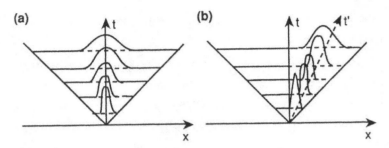

Figure 6. (a) Symmetric random walk ($p = 1/2$). (b) Asymmetric random walk ($p = 3/4$); note that as the center of mass picks up a speed $\beta = p - q$, the rate of spread goes down by a factor of $1 - \beta^2$.

-453-

just as the old system did in the old frame—except that now $\sigma = \sqrt{pqt} = \sqrt{(1 - \beta^2)t/4}$, so that the diffusion will appear to have *slowed down* by a factor of $1 - \beta^2$ (fig. 6(b)).

Intuitively, as some of the resources of the "random walk computer" are shifted toward producing coherent macroscopic motion (uniform motion of the center of mass), *fewer resources will remain available* for the task of producing incoherent motion (diffusion). Thus, we get a slowdown reminiscent of the Lorentz–Fitzgerald "time expansion." In the present situation, however, the slowdown factor is $1 - \beta^2$, related to, but different from, the well-known relativistic factor $\sqrt{1 - \beta^2}$; the transformation that will restore invariance of the dynamics in this case is a Lorentz transformation followed by a scaling of both axes by a further factor of $\sqrt{1 - \beta^2}$.

4.2 A LORENTZ-INVARIANT MODEL OF DIFFUSION

In the above example, when we tried to offset by a *Galilean* coordinate transformation the consequences of a transformation of the dynamical parameter p, we noticed that "proper time," as measured by σ, was not independent of μ. Time as well as space needed to be transformed in order to have the desired dynamical invariance. However, the appropriate transformation was not simply a Lorentz transformation.

But neither were we following the standard procedures of relativity. The fact is, with dynamical *parameters* we are barking up the wrong tree. What relativity says is that a certain kind of transformation of the spacetime coordinates (Lorentz transformation) can always be offset by an appropriate transformation of the dynamical *variables*. We shall now present a lattice-gas model of diffusion that has the same macroscopic phenomenology as the random walk, but is microscopically deterministic and reversible. Unlike the random walk, changes in the macroscopic parameters μ and σ arise in this model from changes in the initial distribution of microscopic states, rather than by tampering with the microscopic *laws*. This model is exactly Lorentz invariant in the continuum limit, i.e., as the lattice spacing λ goes to zero.

Let us consider a one-dimensional cellular automaton having the format of figure 7(a). This is a regular spacetime lattice, with a given spacing λ (lattice units per meter). The arcs represent *signals* traveling at unit speed (the "speed of light"); the nodes represent *events*, i.e., interactions between signals. If one of the possible signal states, denoted by the symbol 0, is interpreted as representing the *vacuum*, the remaining states can be interpreted as *particles* traveling on fixed spacetime tracks (the arcs) and interacting only at certain discrete loci (the nodes). Such a system can be thought of as a *lattice gas* (cf. Hardy, de Pazzis, and Pomeau 1976; Toffoli and Margolus 1987).

Here, we will allow no more than one particle on each track. When two particles collide, each reverses its direction (fig. 7(b)). As long as particles are identical (say, all black), this reversal is indistinguishable from no interaction (fig. 7(c)).

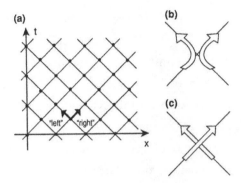

Figure 7. (a) One-dimensional lattice, unfolded over time. The tracks, with slope ±1, indicate potential particle paths; the nodes indicate potential collision loci. (b) Bouncing collision. (c) No-interaction collision.

Now let us paint just one particle red (in which case the reversal *does* make a difference) and study the evolution of its probability distribution $\rho(x; t)$ when both right- and left-going particles are uniformly and independently distributed with linear density (particles per meter) $s = n/\lambda$—where n is the lattice *occupation density* (particles per track).

For fixed s, as $\lambda \to 0$ (continuum limit), $\rho(x; t)$ converges to the solution of the *telegrapher's equation*

$$\partial_{tt}\rho = \partial_{xx}\rho - 2s\,\partial_t\rho. \qquad (17)$$

The latter distribution, in turn, converges to the solution of the diffusion equation

$$\partial_t\rho = \frac{1}{2s}\,\partial_{xx}\rho \qquad (18)$$

in the same circumstances (i.e., $t \to \infty$, $|x| < \sqrt{t}$) as the binomial distribution does. We shall now introduce the freedom to independently vary the densities s_+ and s_- of, respectively, the right- and left-going particles; as a consequence, the center of mass of the red particle's distribution will drift, and its diffusion rate will be affected, too—much as in the asymmetric random walk case. However, this time we have strict Lorentz invariance (in the continuum limit): to every Lorentz transformation of the coordinates, $t, x \mapsto t', x'$, there corresponds a similar linear transformation of the initial conditions, $s_+, s_- \mapsto s'_+, s'_-$, that leaves the form of ρ invariant. (Indeed, the telegrapher's equation is just another form of the Klein–Gordon equation used in relativistic quantum mechanics.)

Lorentz invariance emerges in a similar way for a much more general class of dynamics on a lattice, as explained in Toffoli (1989b); more generally, features qualitatively similar to those of special relativity appear whenever *fixed* computational resources have to be apportioned between producing the inertial motion of a macroscopic object as a whole and producing the internal evolution of the object itself (see Chopard 1990). Thus, we conjecture that special relativity may ultimately be derived from a simpler and more fundamental principle of *conservation of computational resources.*

4.3 GENERAL RELATIVITY

The spacetime lattice in which the particles of the above example move and interact can be thought of as a uniform combinational network—the simplest kind of parallel computer. Recall, however, that Lorentz invariance was achieved in the limit of a vanishingly fine lattice spacing, while holding the density (particles *per meter*) constant. In this limit, then, the occupation number (particles *per track*) goes to zero; this corresponds to a vanishing utilization of the network's computing resources. By the time Lorentz invariance emerges, the model has become useless as a numerical computation scheme.

In an attempt to trade accuracy for computational efficiency, suppose we start backing up a little from the limit, i.e., we consider a network with a fine, but not *infinitely* fine, spacing. As the network becomes coarser, the number of tracks available to the same number of particles decreases, and thus the occupation number increases. When this number significantly departs from zero, the macroscopic dynamics will start deviating from special relativity.

Is this really an unfortunate state of affairs? After all, we know that physics itself starts deviating from special relativity when one dumps more and more matter in the same volume. Are we witnessing the emergence of general relativity? Indeed, the slowdown of the macroscopic evolution brought about, in models of the above kind, by the "crowding" of the computational pathways is strikingly analogous to the proper-time dilation that, in physics, is brought about by the gravitational potential.

~457~

Without more comprehensive models, precise interpretation rules, and quantitative results, any claim that the present approach might have anything to do with modeling general relativity is, of course, premature. But it is legitimate to ask whether fine-grained computation in uniform networks has at least the right kind of internal resources for the task. In other words, is the emergence plausible, in such systems, of a dynamics of spacetime analogous to that described by general relativity? And how could it come about?

Let us start with a metaphor. On a strip of blank punch tape we can record information at a density of, say, ten characters per inch. What if we could only avail ourselves of *used* tape, found in somebody's wastebasket? Knowing the statistics of the previous usage, one can devise appropriate group-encoding techniques and error-correcting codes so as to make such a used tape perfectly adequate for recording new information (Rivest and Shamir 1982)—at a lower density, of course, i.e., up to the maximum density allowed by Shannon's theorems for a noisy channel. The *proper length* of the tape, defined in terms of how many characters *we* can record on it, will be less than that of blank tape, by a factor that will depend on how heavy the original usage was. If the tape is sufficiently long, its statistical properties may significantly vary from place to place, and we may want to adapt our encoding strategy to the local statistics—yielding a *proper-length* metric that varies from place to place.

Let us extend the above metaphor from the domain of information *statics* to that of information *dynamics*. Consider, for example, a programmable gate array having a nominal capacity of, say, 10,000 gates. An inventor designs a clever arcade game that takes full advantage of the chip's "computing capacity," and asks the VLSI factory to produce a million copies of it. The game turns out to be a flop, and the programmed chips get thrown in the wastebasket. What is the effective "computing capacity" of these chips from the viewpoint of the penniless but undaunted hacker that finds them? How many of these chips would he have to put together in order to construct *his own* arcade game, and how many clock cycles of the original chip would he have to string together to achieve a usable clock cycle for his game? What in the new game is simply the toggling of a flip–flop may correspond, in the underlying original game, to the destruction of a stellar empire. For the new user, *proper time* will be measured in terms of how fast the evolution of *his* game can be made to proceed.

For a macroscopic scavenger, the individual hole positions in a punched tape or the individual gates in an electronic circuit blend into a continuum, locally characterized by a certain effective density of information-storage capacity and a certain effective density of information-processing capacity. These densities reflect the constraints that the local "degree of congestion" of the computing resources set on any "further incremental usage" of these resources. Thus, if *length* and *time* measure, respectively, the effective information-storage and information-processing capacities available to macroscopic epiphenomena, a metric and a dynamics of *curved* spacetime naturally emerge out of a *flat*, uniform computing network.

4.4 CONCLUSIONS

Quantitative features of special relativity and at least qualitative features of general relativity emerge quite naturally as epiphenomena of very simple computing networks. Thus, relativity appears to be of the right form to be an emergent property, whether or not *that* is the way it comes about in physics.

5. General Conclusions

Many of what are regarded as the most fundamental features of physics happen to have the right form to be emergent features of a much simpler fine-grained dynamics.[4]

A century and a half ago, most people were happy with the idea that the cell was a bag of undifferentiated "protoplasm" governed by some irreducible "vital force." The behavior of the cell was obviously very rich, but few people dared to ascribe it to much finer-grained internal machinery, explicitly built according to immensely detailed blueprints.

Today we know for sure about the existence of such machinery and such blueprints. Besides molecular genetics, chemistry and nuclear physics provide further case histories where complex behavior was successfully reduced to simpler primitives on a grain a few orders of magnitude finer.

---

[4]Even *invertibility*—perhaps the most strongly held feature of microscopic physics—can quite naturally emerge out of an underlying non-invertible dynamics. We are going to discuss this topic in a separate paper.

For a physicist, the possibility of explanation by reduction to simpler, smaller structures is of course one of the first things that comes to mind. The point of this paper is that one should look for such possibility not only to explain specific phenomenology, but also to re-examine those general principles that are so familiar that no "explanation" seems to be needed. ✈

Acknowledgments

This research was supported in part by the Defense Advanced Research Projects Agency (N00014-89-J-1988) and in part by the National Science Foundation (8618002-IRI).

REFERENCES

Arnold, V. 1978. *Mathematical Methods of Classical Mechanics*. Berlin, Germany: Springer-Verlag.

Chopard, B. 1990. "A Cellular Automata Model of Large-Scale Moving Objects." *Journal of Physics A: Mathematical and General* 23 (10): 1671–87. doi:10.1088/0305-4470/23/10/010.

Hardy, J., O. de Pazzis, and Y. Pomeau. 1976. "Molecular Dynamics of a Classical Lattice Gas: Transport Properties and Time Correlation Functions." *Physical Review A* 13 (5): 1949–61. doi:10.1103/physreva.13.1949.

Kuhn, T. 1970. *The Structure of Scientific Revolutions*. 2nd ed. Chicago, IL: University of Chicago Press.

Margolus, N. 1988. *Physics and Computation*. Technical report. Tech. Rep. MIT/LCS/TR-415, MIT Laboratory for Computer Science.

Rivest, R. L., and A. Shamir. 1982. "How to Reuse a "Write-Once" Memory." *Information and Control* 55 (1–3): 1–19. doi:10.1016/S0019-9958(82)90344-8.

Schroeder, M. 1986. *Number Theory in Science and Communication*. 2nd enlarged ed. Berlin, Germany: Springer-Verlag.

Toffoli, T. 1989a. *Analytical Mechanics from Statistics:* $T = dS/dE$ *Holds for Almost Any System.* Tech. Memo MIT/LCS/TM-407, MIT Laboratory for Computer Science.

————. 1989b. "Four Topics in Lattice Gases: Ergodicity; Relativity; Information Glow; and Reule Compression for Parallel Lattice-Gas Machines." In *Discrete Kinetic Theory, Lattice Gas Dynamics, and Foundaitons of Hydrodynamics,* edited by R. Monaco, 343–54. Singapore: World Scientific.

Toffoli, T., and N. Margolus. 1987. *Cellular Automata Machines—A New Environment for Modeling.* Cambridge, MA: MIT Press.

6K

INTERMITTENT FLUCTUATIONS
AND COMPLEXITY

Xiao-Jing Wang, University of Texas, Austin

I. Introduction

We shall summarize here succinctly some recent progress in our understanding of intermittent phenomena in physics. Intermittency often refers to random, strong deviations from regular or smooth behavior. Consider, for instance, an iterative dynamical system (fig. 1)

$$x_{n+1} = f(x_n) = x_n + x_n^z \quad (\text{mod } 1). \tag{1}$$

For $z = 3$, if we start with an initial condition $x_0 = 0.001$, then $x_0^z = 10^{-9}$, and the system would remain near the origin for millions of time units, before suddenly turning into a burst of irregular oscillations with considerable amplitude. In such a temporal evolution with long quiescent periods spontaneously interspersed by random events, an observable can be "almost surely constant in every prescribed finite span of time," as Mandelbrot (1967) once put it, "but it almost surely varies sometime."

Equation (1) is called the Manneville–Pomeau map (Bergé, Pomeau, and Vidal 1984), at the transition point from a periodic state (the fixed point $x = 0$) to a chaotic state. It played an important role in the study of the onset of turbulence. To describe this intermittent dynamics, two quantities are most relevant: One is the fraction of time during which the output is irregular ("turbulent") or when the signal is larger than a

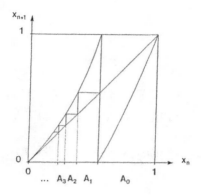

Figure 1. The Manneville–Pomeau map, with a countable partition of the phase space $(0, 1)$.

threshold, say, $x \in A_0 = (c, 1)$ with $1 = c + c^z$. Thus, the "turbulent time" may be related to the number N_n of recurrences to the cell A_0 during a time span n, and the "laminar time" is $n - N_n$.

The other quantity (perhaps even more important) is the Lyapunov exponent given as

$$\lambda = \lim_{n \to \infty} \frac{\Lambda_n}{n}, \quad \text{with} \quad \Lambda_n(x) = \sum_{k=0}^{n-1} \log |f'(f^k(x))|. \quad (2)$$

Thus, if $\lambda > 0$, there is an exponential sensitivity to initial conditions

$$\delta x_n \simeq \prod_{k=0}^{n-1} |f'(x_k)| \, \delta x_0 \simeq e^{\lambda n} \delta x_0, \quad (3)$$

and the behavior is then said to be chaotic. An *entropy* is also well defined for dynamical systems, thanks to Kolmogorov and Sinai (Eckmann and Ruelle 1985). The idea is that a *deterministic* chaotic system admits a *discrete* generating partition of its phase space, resulting in an exactly equivalent *stochastic* process with discrete (finite or denumerable) states. The Kolmogorov–Sinai entropy h_{KS} is then equal to the Shannon entropy per time unit of the associated stochastic process. For chaotic *attractors* in one-dimensional mappings, h_{KS} coincides with λ.

Now, for intermittent cases, the regular (or "laminar") phases may be so prevailing that the irregular (or "turbulent") oscillations could occur only in a subset of the time axis with zero measure. This may happen if

$$\Lambda_n \sim n^\nu \quad \text{for} \quad 0 < \nu < 1, \tag{4}$$

which implies $\Lambda_n/n \to 0$, and the dynamic stability is *stretched exponential* rather than exponential. This kind of behavior was called "sporadic" (Gaspard and Wang 1988). We shall show that it represents a special class of intermittent systems, with the *algorithmic complexity* of Kolmogorov and Chaitin of a form intermediate to predictable and random cases.

A fully disordered or random system is sometimes perceived as simple rather than complex, mostly when its *fluctuations* appear small and inconspicuous. On the other hand, the intermittency is generally characterized by *abnormal* fluctuations and $1/f$-noise-like power spectrum (Ben-Mizrachi et al. 1985; Geisel, Nierwetberg, and Zacherl 1985), and *local* fluctuations around the mean of an observable may obey a Lévy, rather than a Gauss, distribution. More complete information is provided by knowledge about *large fluctuations*, using the thermodynamic formalism of Sinai, Ruelle, and Bowen for dynamical systems (Paladin and Vulpiani 1987). The Sinai–Ruelle–Bowen theory furnished a rigorous connection between dynamical systems and equilibrium statistical mechanics in one dimension (on the time axis). In this framework abnormal behaviors of large deviations in equation (1) are to be treated as a problem of phase transition in its statistical mechanical counterpart (Wang 1989a, 1989b).

To sum up, one can look at the mean, local fluctuations, large deviations, and equilibrium statistical mechanics in order to achieve increasingly detailed descriptions of intermittent systems. Furthermore, one can also study a system's unusual *non-equilibrium* properties, following a suggestion of Nicolis and Nicolis (1988, 1990). In what follows we shall take equation (1) for a case study to illustrate how each of these levels of consideration leads to insights into new aspects of fluctuations and complexity of the intermittent processes. The next two sections are devoted to the equilibrium properties of equation (1), and then non-equilibrium is discussed in the following section. Some other examples with close analogy to the intermittent system will be mentioned in the final section, including a discrete one-dimensional model of Anderson localization in disordered matter.

~466~

Initially observed in fluid turbulence, intermittency has more recently been evidenced in other physical systems as diverse as the large-scale structure of the universe and the hadronic multiparticle production in high-energy physics. Little is known, as yet, beyond the phenomenology of these spatially extended processes.

II. Algorithmic Complexity of Sporadic Behavior

This section is largely based on Gaspard and Wang (1988). Let

$$S = (s_0 s_1 s_2 s_3 \ldots) \qquad (5)$$

be a string of integers or symbols. The algorithmic complexity (Chaitin 1987; Kolmogorov 1983) $K(S_n)$ of the string S_n composed of the n first symbols of equation (5) is defined as the binary length of the shortest possible program able to reconstruct the string S_n on a universal machine. Thus, any sequence that admits a finite program to generate it can be reproduced by specifying this program and the length n of the sequence, so that

$$K(S_n) \sim \log_2 n. \qquad (6)$$

This "predictable" class of strings includes periodic ones (for which it suffices to specify the pattern of one period in the program), as well as a large set of aperiodic ones. An often-cited example is the digital expansion of the mathematical constant Pi, $\pi = 3.141592\ldots$, for which a convergent series representation exists (e.g., in the sum of S. Ramanujan's series each successive term adds roughly eight correct digits).

On the other hand, for a random sequence where no regularity can be found, one could only "copy" the whole string, bit by bit, so that

$$K(S_n) \sim n. \tag{7}$$

In chaotic systems with a positive entropy h_{KS}, almost all trajectories are random, in the sense that

$$\lim_{n\to\infty} \frac{1}{n} K(S_n) = h_{\text{KS}} > 0. \tag{8}$$

To estimate the algorithmic complexity for the intermittent system given in equation (1), we observe first that with the countable partition of $(0, 1)$ as shown in figure 1, the system can be approximated by a Markov chain with denumerable states. The transition matrix takes the form

$$
\begin{pmatrix}
p_{00} & p_{01} & p_{02} & p_{03} & \cdots \\
1 & 0 & 0 & 0 & \cdots \\
0 & 1 & 0 & 0 & \cdots \\
0 & 0 & 1 & 0 & \cdots \\
\vdots & \vdots & \vdots & \vdots & \ddots
\end{pmatrix}, \tag{9}
$$

with the transition probability p_{0n} and the invariant measure μ_n satisfying

$$p_{0n} \sim_{n\to\infty} \frac{1}{n^{1+\alpha}} \quad \text{and} \quad \mu(A_n) \sim_{n\to\infty} \frac{1}{n^\alpha}, \tag{10}$$

where $\alpha = 1/(z - 1)$.

An orbit from this process would be

$$S = 543210210087654321032109876\dots \tag{11}$$

Because of the predictable nature of the "laminar" phases, S is uniquely recovered from a shorter string

$$R = 520839\dots, \tag{12}$$

~468~ which consists solely of the symbols immediately after each "turbulent" state (the state A_0). S is therefore compressible. Assume that a finite string of length n has N_n recurrences to A_n; then the complexity of S_n may be estimated as

$$K(S_n) = \sum_{i=1}^{N_n} \log_2 s_{k_i} \tag{13}$$

with $s_{k_{i-1}} = 0$. Using the theory of recurrent events, one can then show that

$$E(K(S_n)) \sim \begin{cases} n & \text{if } 2/3 < z < 2; \\ n^{\frac{1}{z-1}} & \text{if } 2 < z; \\ \dfrac{n}{\log n} & \text{if } z = 2. \end{cases} \tag{14}$$

Therefore, when $2 \leq z$, the behavior is intermediate between the predictable and random cases, in the sense that the algorithmic complexity is

$$K_n \sim n^{\nu_0}(\log n)^{\nu_1} \quad \text{with} \quad 0 < \nu_0 < 1 \quad \text{or} \quad \nu_0 = 1, \nu_1 < 0. \tag{15}$$

Although equation (15) seems to betoken a continuous spectrum with periodicity ($\nu_0 = 0$) and randomness ($\nu_0 = 1$) at the two extremities, these *sporadic* behaviors stand in qualitative contrast with both totally ordered and completely disordered ones.

III. Abnormal Fluctuations and Phase Transition à la Fisher

A prominent property of the intermittent dynamics in equation (1) is that long "laminar" phases entail a long-range time correlation in power law (Wang 1989a, 1989b),

$$\phi_n \sim \begin{cases} n^{-(\alpha-1)} & \text{if } 1 < z < 2 \quad (1 < \alpha); \\ n^{-2(1-\alpha)} & \text{if } 2 < z \quad\quad (0 < \alpha < 1). \end{cases} \quad (16)$$

~469~

The fluctuations near the mean value of an observable may not be Gaussian, e.g., the variance of fluctuations for the turbulent time N_n is (Gaspard and Wang 1988)

$$\text{Var}(N_n) \sim \begin{cases} n & \text{if } 1 < z < 2/3; \\ n^{3-\alpha} & \text{if } 2/3 < z < 2; \\ n^{2\alpha} & \text{if } 2 < z. \end{cases} \quad (17)$$

For $z < 2/3$, there exists a central limit theorem asserting the Gaussian character of the fluctuations. For $2/3 < z$, on the other hand, they obey a generalized limit theorem involving the Lévy stable distribution $g_\alpha(x)$ with $0 < \alpha \le 2$ (see Montroll and West 1987). ($g_{\alpha=2}(x)$ is the familiar Gauss law. A Lévy distribution with $\alpha < 2$ enjoys a cognate genericity as a Gauss distribution for sums of independent random variables with a common distribution, only now this latter distribution has an infinite second moment.)

In both cases, the correlation function is a power law. This tells us that local fluctuations are not sufficient to characterize the abnormality of the system. In fact, a central limit theorem is concerned with fluctuations of the form

$$\frac{1}{n}\sum_{k=0}^{n-1} g(x_k) - E(g) \le \frac{\sigma\varsigma}{\sqrt{n}}, \quad -\infty < \varsigma < +\infty, \quad (18a)$$

where $E(g)$ stands for the mean value of an observable $g(x)$. Instead of (18a), one can consider large deviations, i.e.,

$$\frac{1}{n}\sum_{k=0}^{n-1} g(x_k) \in (\Lambda, \Lambda + d\Lambda) \qquad (18b)$$

for all possible values Λ, not necessarily near its mean value. If $g(x) = \log|f'(x)|$, then the left side of equation (18b) tends to the Lyapunov exponent λ, and one is dealing with

$$U_n(s_0 s_1 s_2 \ldots s_{n-1}) = \inf_{x \in I(s_0 s_1 s_2 \ldots s_{n-1})} \sum_{k=0}^{n-1} \ln|f'(f^{(k)}(x))|,$$

$$(19)$$

where $I(s_0 s_1 s_2 \ldots s_{n-i})$ is the cell in the phase space coded by $(s_0 s_1 s_2 \ldots s_{n-i})$.

A fundamental result in the Sinai–Ruelle–Bowen theory states that the invariant measure of a dynamical system is given by

$$\text{Prob}(s_0 s_1 s_2 \ldots s_{n-1}) \sim \exp(-U_n(s_0 s_1 s_2 \ldots s_{n-1})), \qquad (20)$$

which takes a similar form to a Gibbs state in equilibrium statistical mechanics, with the inverse of a temperature $\beta = 1$. This expression is to be especially appreciated because for dissipative dynamical systems the invariant measure is unknown a priori, unlike for Hamiltonian systems. Following this line of thinking, one is led to seek a mapping of the dynamical system into a statistical mechanical system (on a one-dimensional lattice), with the Hamiltonian given by equation (19). Λ then plays the role of energy, and its thermodynamic conjugate quantity is β (the formally introduced parameter β is interpreted physically in this way). Similarly, if one identifies the "laminar" (respectively "turbulent") state with the presence (respectively absence) of a particle on a given lattice site, $n - N_n$ represents the number of particles in a large albeit finite lattice. Hence, a density and a chemical potential can be accordingly introduced. The advantage of these identifications is to embed the problem of large deviations into the framework of equilibrium statistical mechanics.

A detailed analysis of the intermittent system described by equation (1) (or rather a piecewise-linear approximation of it) is carried out in Wang (1989a, 1989b). Let us call attention to some main conclusions therefrom. It was found that the system's statistical mechanical counterpart bears a close analogy to a droplet model of condensation proposed by Fisher (1967) about twenty-five years ago. The clusters of "laminar" states are similar to the clusters of particles (the droplets) in Fisher's model of gas–liquid phase transition; there are many-body interactions within each cluster, which results in a surface energy of logarithmic type.

On the pressure–temperature plane, there are two thermodynamic phases separated by a critical line. They are the chaotic ("gas") and periodic ("condensed") states, and the intermittent state is located on the codimension-one critical curve of phase transition. This is true for all $1 < z$, regardless of whether the local fluctuations are Gaussian or not. Therefore, the abnormal large fluctuations may be detected as a phase transition of the associated statistical mechanical system. The identification of the interaction potential and the types of resulting critical phenomena constitute a finest characterization and universal classification of such intermittent dynamical systems.

IV. Approach to Equilibrium is Not Exponential

The previous sections are concerned mainly with stationary or *equilibrium* properties of the intermittent system. In the present section we shall mention briefly certain *non-equilibrium* aspects of dynamical systems. The discussion is inspired by the recent work of Nicolis and Nicolis (1988, 1990), in which a *master equation* approach to deterministic chaos has been advanced. A central question to be addressed is this: Given a non-equilibrium initial distribution of points on an attractor, how will it converge to the equilibrium distribution (i.e., the invariant measure)?

In the case of our countable Markov model of intermittency, the answer is surprisingly straightforward. Indeed, a theorem due to D. G. Kendall (1960) states that *for any irreducible and aperiodic Markov chain, the convergence is exponential if and only if there is a state i such that*

$$\sum_{n=1}^{\infty} f_{ii}^n s^n, \qquad (21)$$

where f_{ii}^n denotes the probability of first recurrence at time n of the state i, has a radius of convergence greater than unity.

Applying this theorem to the state A_0 in our case, with the probability of first recurrence given by $p_{0(n-i)}$, $p_{0n} \sim 1/n^{-(1+\alpha)}$ (eq. 10) immediately implies that the radius of convergence of the sum in equation (21) is unity. Hence, one concludes that the convergence to equilibrium is slower than any exponential law.

Let us indicate why this might be expected from the viewpoint of the spectral properties of the transition matrix in equation (9). According to the Perron–Frobenius theory (see Seneta 1973), a finite non-negative matrix, say, (a_{ij}), $i, j = 1, 2, \ldots, m$, possesses a unique maximum eigenvalue λ_0 such that

$$\min_i \sum_{j=1}^{m} a_{ij} \le \lambda_0 \le \max_i \sum_{j=1}^{m} a_{ij}. \qquad (22)$$

For the transition matrix of a finite Markov chain, this sum is $\sum_j p_{ij} \equiv 1$. Thus, $\lambda_0 = 1$, and all the other eigenvalues have a modulus less than 1. Any initial distribution will then be projected onto the (non-negative) eigenvector associated with

λ_0, i.e., the invariant measure, and all the other components vanish in an exponential fashion.

Now, for a countable Markov chain, the transition matrix has denumerably infinite eigenvalues, so that the eigenvalue $\lambda_0 = 1$ may be approached arbitrarily by other eigenvalues. This is indeed the case for the model of intermittency. Let us sketch the argument. Truncating the transition matrix in equation (9) up to n, one obtains a finite matrix $\mathbf{W}_n$ of which the characteristic equation is

$$F_n(\lambda) = \lambda^n - p_{00}\lambda^{n-1} - p_{01}\lambda^{n-2} - \cdots - p_{0(n-2)}\lambda - p_{0(n-1)} = 0$$

(23)

- 473 -

with $\lim_{n\to\infty} F_n(\lambda_0 = 1) = 0$. It follows from the Perron–Frobenius theory that all the roots of equation (23) are confined inside the unit disc.

Considering $F_n(x)$ in equation (23) as the partial sum of an infinite series, one can readily see that the radius of convergence of this series is 1. Now it is useful to invoke a remarkable theorem in analysis, due to R. Jentzsch (Titchmarsh 1932), which asserts that *for every power series, every point on the circle of convergence is a limit-point of zeros of partial sums.* Hence, the $\lambda_0 = 1$ of the *countable* chain is a limit-point of roots of equation (23). On the other hand, it is reasonable to assume that as $n \to \infty$, every such root is arbitrarily close to one of the true eigenvalues of the infinite matrix in equation (9). One concludes therefore that $\lambda_0 = 1$ is not isolated. This suggests, although it does not ensure, that the approach to equilibrium may not have an exponential rate.

V. Concluding Remarks

There is an increasing number of systems to which the present work appears relevant. Such a case cited in Gaspard and Wang (1988) is the Markov model defined on a tree that was proposed by J. Meiss and E. Ott for Hamiltonian chaos, in the presence of a hierarchy of "cantori." Another example is a model of abnormal diffusion in resistively shunted Josephson junctions, which takes a form similar to equation (1) (Geisel, Nierwetberg, and Zacherl 1985). Perhaps more surprisingly, it has been noticed (Bouchard and Le Doussal 1986) that a discrete one-dimensional model of Anderson localization seems also somewhat akin to the intermittent map equation (1). Let us end this paper with a few remarks on this intriguing finding.

The one-dimensional Anderson model is a one-dimensional discrete Schrödinger equation with a random potential $\{V_n\}$,

$$\Phi_{n+1} - 2\Phi_n + \Phi_{n-1} = (E - V_n)\Phi_n. \tag{24}$$

Letting $R_n = \Phi_n/\Phi_{n-1}$, equation 24 can be rewritten as

$$R_{n+1} = (2 + E) - V_n - 1/R_n. \tag{25}$$

All the states within the pure energy band $[-4, 0]$ are localized in the presence of a random potential $\{V_n\}$, with the inverse localization length directly given by the Lyapunov exponent of the map in equation (25). For $V_n \equiv 0$, the map in equation (25) is displayed in figure 2, with $\log|f'(R_n)| = -2\log|R_n|$. *Locally* around $x = 1$, the mapping at the band edge $E = 0$ looks the same as figure 1 with $z = 2$, and it is of great interest to recognize such a resemblance between Anderson localization and intermittency. There are, however, notable differences which perhaps should not be overlooked. In contrast to intermittent systems, here the mapping is invertible, and the Lyapunov exponent is always zero even for $0 < E$, if the stochastic term is absent (obviously, no localization is possible without random potential; see Derrida and Gardner 1984). Besides, due to the *linear* character of equation (24), the Thouless (1972) formula asserts a direct

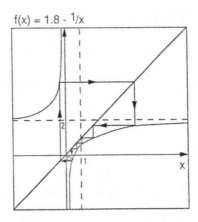

f(x) = 1.8 - ^{1}/x

Figure 2. The map equation (25) in the absence of the noise term. It arises from a one-dimensional Anderson model.

relationship between the inverse localization length and the integrated density of states. Interpreted in terms of the dynamics in equation (25), the former is the Lyapunov exponent while the latter, being related to the number of nodes of the wave function, is also the number of times that R_n is negative in the lattice, hence assimilable to the "turbulent time" (cf. fig. 2). Such a "dispersion relation" between the Lyapunov exponent and the turbulent time, however, does *not* seem to exist for the *nonlinear* intermittent system: it would lead to the erroneous conclusion that the Lyapunov exponent in the latter case is also identically zero in the absence of noise.

On the other hand, the thermodynamic description of equation (1) (Wang 1989b) does provide a connection between the entropy S (here equivalent to the Lyapunov exponent) and the density of the laminar phase ρ (i.e., one minus the fraction of the turbulent time). Both S and ρ are functions of the two thermodynamic variables β (inverse of temperature) and μ (chemical potential), and are related one to the other by the fundamental equation of thermodynamics,

$$\beta p(\beta, \mu) = \beta \rho \mu - \beta \Lambda + S. \qquad (26)$$

Hence, we have a *thermodynamic*, rather than dispersion, relation in the case of such nonlinear intermittent processes. ✦

Acknowledgment

It is a pleasure to thank warmly Professor G. Nicolis for his continuous help, encouragement, and fruitful correspondence. This work was partly supported by the Department of Energy under contract number DE-AS05-81ER10947. Sincere thanks are also due to the Center for Statistical Mechanics at the University of Texas for financial support of my attendance at the Santa Fe Institute Workshop.

REFERENCES

Ben-Mizrachi, A., I. Procaccia, N. Rosenberg, A. Schmidt, and H. G. Schuster. 1985. "Real and Apparent Divergencies in Low-Frequency Spectra of Nonlinear Dynamical Systems." *Physical Review A* 31 (3): 1830–40. doi:10.1103/physreva.31.1830.

Bergé, P., Y. Pomeau, and C. Vidal. 1984. *L'ordre dans le Chaos*. Paris, France: Herman.

Bouchard, J. P., and P. Le Doussal. 1986. "Intermittency in Random Optical Layers at Total Reflection." *Journal of Physics A: Mathematical and General* 19 (5): 797–810. doi:10.1088/0305-4470/19/5/033.

Chaitin, G. 1987. *Algorithmic Information Theory*. Cambridge, UK: Cambridge University Press.

Derrida, B., and E. Gardner. 1984. "Lyapounov Exponent of the One-Dimensional Anderson Model: Weak Disorder Expansions." *Journal de Physique* 45 (8): 1283–95. doi:10.1051/jphys: 01984004508012830o.

Eckmann, J.-P., and D. Ruelle. 1985. "Ergodic Theory of Chaos and Strange Attractors." *Reviews of Modern Physics* 57 (3): 617–56. doi:10.1103/RevModPhys.57.617.

Fisher, M. E. 1967. "The Theory of Condensation and the Critical Point." *Physics Physique Fizika* 3 (5): 255–83. doi:10.1103/physicsphysiquefizika.3.255.

Gaspard, P., and X.- J. Wang. 1988. "Sporadicity: Between Periodic and Chaotic Dynamical Behaviors." *Proceedings of the National Academy of Sciences* 85 (13): 4591–95. doi:10.1073/pnas.85.13.4591.

Geisel, T., J. Nierwetberg, and A. Zacherl. 1985. "Accelerated Diffusion in Josephson Junctions and Related Chaotic Systems." *Physical Review Letters* 54 (7): 616–19. doi:10 . 1103 / physrevlett.54.616.

Kendall, D. G. 1960. "Geometric Ergodicity and the Theory of Queues." In *Mathematical Methods in the Social Sciences,* edited by K. J. Arrow, S. Karlin, and P. Suppes, 176–95. Palo Alto, CA: Stanford University Press.

Kolmogorov, A. N. 1983. "Combinatorial Foundations of Information Theory and the Calculus of Probabilities." *Russian Mathematical Surveys* 38 (4): 29–40. doi:10 . 1070 / rm1983v038n04abeh004203.

Mandelbrot, B. B. 1967. "Sporadic Random Functions and Conditional Spectral Analysis: Self-Similar Examples and Limits." In *Proceedings of the Fifth Berkeley Symposium on Mathematical Statistics and Probability,* edited by L. LeCam and J. Neyman, 155–79. Berkeley, CA: University of California Press.

Montroll, E. W., and B. J. West. 1987. "On an Enriched Collection of Stochastic Processes." In *Fluctuation Phenomena,* revised, edited by E. W. Montroll and J. L. Lebowitz, 61–206. Amsterdam, Netherlands: North-Holland.

Nicolis, G., and C. Nicolis. 1988. "Master-Equation Approach to Deterministic Chaos." *Physical Review A* 38 (1): 427–33. doi:10.1103/physreva.38.427.

———. 1990. "Chaotic Dynamics, Markovian Coarse-Graining and Information." *Physica A: Statistical Mechanics and its Applications* 163 (1): 215–31. doi:10.1016/0378-4371(90)90331-1.

Paladin, G., and A. Vulpiani. 1987. "Anomalous Scaling Laws in Multifractal Objects." *Physics Reports* 156 (4): 147–225. doi:10.1016/0370-1573(87)90110-4.

Seneta, E. 1973. *Non-Negative Matrices.* New York, NY: John Wiley & Sons.

Thouless, D. J. 1972. "A Relation between the Density of States and Range of Localization for One-Dimensional Random Systems." *Journal of Physics C: Solid State Physics* 5 (1): 77–81. doi:10.1088/0022-3719/5/1/010.

Titchmarsh, E. C. 1932. *The Theory of Functions.* Oxford, UK: Oxford University Press.

Wang, X.-J. 1989a. "Abnormal Fluctuations and Thermodynamic Phase Transitions in Dynamical Systems." *Physical Review A* 39 (6): 3214–17. doi:10.1103/physreva.39.3214.

———. 1989b. "Statistical Physics of Temporal Intermittency." *Physical Review A* 40 (11): 6647–61. doi:10.1103/physreva.40.6647.

69

INFORMATION PROCESSING IN VISUAL PERCEPTION

A. Zee, Institute for Theoretical Physics

We study various quantitative issues in the theory of visual perception

centering around two fundamental questions: How perceptive are we?

And how are we as perceptive as we are? We outline a research program

to answer these questions.

The problem of understanding visual perception (Albrecht 1982; Levine and Shefner 1981; Marr 1982) surely ranks as one of the outstanding scientific problems of our time. Of all the problems that bear upon the ultimate mystery of understanding how the brain works, the problem of visual perception is perhaps the one most amenable to rigorous experimental study and quantitative theory making. I would like to review briefly some work I have done with W. Bialek.

The questions and issues we addressed can be divided into three areas. We attempted to quantify some of the issues involved, with the hope of sharpening these issues and of bringing some ideas into direct confrontation with experiments.

1. We humans in general, and physicists in particular, tend to think that our brains function extremely well, especially at processing visual information. In truth, however, a quantitative and even semi-quantitative

measure of the ability of the brain to process information sent to it by the sensory system is lacking.

We simply do not know how well the brain tackles computational problems of varying degrees of complexity.

How perceptive are we? Suppose that it could be established that the visual system performs optimally. Then we can go on to ask what sort of computation is necessary in order to achieve this level of performance, and to explore in detail the types of design that may be required.

2. In the perception literature, a number of models of visual processing have been proposed. We raise the question of how such models can be falsified by experiment. Obviously, it is of prime importance to progress from qualitative discussions to quantitative tests, in order to determine whether a given model is in fact viable. We try to compute the performance that the visual system is capable of according to each of these models. If this performance falls significantly below the experimentally measured performance, then clearly the model can be ruled out. A large class of perceptual models corresponds to the steepest descent or mean field approximation in our formulation. An important issue is whether this approximation is adequate in explaining human performance.

Performance

OPTIMAL PERFORMANCE

To discuss how well the visual system performs, we must immediately raise an obvious question: What are we to compare the performance of the visual system with? The only natural standard, it appears to us, is the optimal performance allowed by information theory, that is, the performance attainable if every bit of information received by the visual system is used.

First, we must choose a "naturalistic" task well suited to the visual system but which also allows a precise mathematical formulation amenable to rigorous analysis. Since we suspect that the visual system can in fact perform at or near optimum, we also want the task to be computationally difficult. We chose the discrimination between patterns with noise and distortion added.[1]

~ 481 ~

More precisely, we propose an experiment in which the subject is first acquainted with two patterns, described by $\phi_0(x)$ and $\phi_1(x)$. Here x denotes the coordinates of the two-dimensional visual field. A black and white pattern is described by a scalar field $\phi(x)$ where $\phi(x)$ is equal to the contrast, that is, the logarithm of the intensity of the pattern at the point x.

For each trial, the experimenter chooses either ϕ_0 or ϕ_1, with equal probability, say. Suppose ϕ_0 is chosen. Then the pattern is distorted and obscured with noise so that the subject actually sees $\phi(x) = \phi_0(y(x)) + \psi(x)$. Here $x \to y(x)$ defines an arbitrary one-to-one mapping of the plane onto itself. The noise $\psi(x)$ is taken for simplicity to be Gaussian and white. Thus, the conditional probability of seeing $\phi(x)$ were ϕ_0 chosen is given by $P(\phi \,|\, \phi_0) = (1/Z) \int Dy \; e^{-W(y) - \beta \int d^2x [\phi(x) - \phi_0(y(x))]^2}$. The functional $W(y)$ should favor gentle distortions, for which $y(x) \sim x$. More on W later. Here Z is a normalization factor required by $\int \mathcal{D}\phi \; P(\phi \,|\, \phi_0) = 1$. Henceforth, we will often neglect to write the normalization factor. Evidently, a probability $P(\phi \,|\, \phi_0)$ can also be defined by substitution. The

---

[1]Experiments similar to (but simpler than) the ones proposed here have been done by Barlow (1982, 1980). Also see Bialek and Zee (1987, 1990b).

subject is to decide whether the pattern seen corresponds to ϕ_0 or ϕ_1.

The patterns ϕ_0 and ϕ_1 should be abstract so as to eliminate any possible biological or cultural bias, such as those associated with our finely developed ability to recognize human faces. We are interested in perception rather than cognition.

DISCRIMINABILITY

The information-theoretic optimal performance can then be computed according to standard signal detection theory. A particularly relevant quantity is the discriminability. Define the discriminant as $\lambda(\phi; \phi_0, \phi_1) = \log[P(\phi \mid \phi_0)/P(\phi \mid \phi_1)]$. With this definition, the discriminant is positive when the probability $P(\phi \mid \phi_0)$ is larger than $P(\phi \mid \phi_1)$ and negative when the opposite holds. (The logarithmic form for the discriminant is chosen for convenience. Some other monotonic function of the ratio of the two probabilities $P(\phi \mid \phi_0)$ and $P(\phi \mid \phi_1)$ may serve equally well.) It can be shown that optimal discrimination is accomplished by maximum likelihood. In plain English, the optimal strategy is to identify the pattern as ϕ_0 if $\lambda(\phi; \phi_0, \phi_1)$ is positive, and as ϕ_1 if $\lambda(\phi; \phi_0, \phi_1)$ is negative. This is, of course, precisely the strategy that any sensible person capable of knowing λ will adopt. Having seen the image $\phi(x)$, we have to decide whether it is more likely that the image "came" from $\phi_0(x)$ or from $\phi_1(x)$. (Thus, the experiment implies a "learning phase" in which the subject tries to "figure out" the relevant probability distributions. We are interested in the performance reached after learning. This, of course, accounts for our insistence on "naturalistic" tasks, for which the necessary learning has already been accomplished through eons of evolution.)

The probability distribution of λ if ϕ_0 is chosen is defined by $\mathcal{P}(\lambda \mid \phi_0; \phi_0 \text{ vs. } \phi_1) = \int D\phi \, \delta(\lambda(\phi; \phi_0, \phi_1) - \lambda) \times P(\phi \mid \phi_0)$. Similarly, $(\lambda \mid \phi_1; \phi_0 \text{ vs. } \phi_1)$ can be defined. The discriminability, conventionally called $(d')^2$, is defined as

$$(d')^2 = \frac{(\langle \lambda \rangle_0 - \langle \lambda \rangle_1)^2}{\frac{1}{2}[\langle (\delta\lambda)^2 \rangle_0 + \langle (\delta\lambda)^2 \rangle_1]},$$

where the subscript $i = 0$ or 1 indicates that the corresponding expectation value should be taken in the distribution $\mathcal{P}(\lambda \mid \phi_0;$ ϕ_0 vs. ϕ_1) or $\mathcal{P}(\lambda \mid \phi_1; \phi_0$ vs. ϕ_1) respectively. The meaning of $(d')^2$ is obvious: it measures the overlap between the two probability distributions when the two distributions are bell shaped. As the name suggests, the discriminability $(d')^2$ limits the extent to which one can discriminate between ϕ_0 and ϕ_1. We are generally interested in the regime $(d')^2 \sim 0$ where the visual discrimination task is highly "confusing." When the distributions are bell shaped, the discriminant can obviously be related to the percentage of correct guesses. Incidentally, the discriminant $(d')^2$, rather than some other more-or-less equivalent quantity, is used because, being formed of "naturally occurring" expectation values, it can be readily computed for certain simple problems and because experimentalists in this field typically quote their observations in terms of $(d')^2$. The discriminant provides a convenient summary of the information contained in the two $\mathcal{P}(\lambda)$s.

FIELD THEORY AND STATISTICAL MECHANICS

As is well known, quantum field theory and statistical mechanics can both be described by functional integrals. Thus, the considerable body of knowledge accumulated about two-dimensional field theories and statistical mechanical systems may be brought to bear on the theoretical problem of determining the various probability distributions and $(d')^2$.

In the last two decades or so, studies in quantum field theory and statistical mechanics have revealed that apparently somewhat simple systems can exhibit exceedingly intricate collective behavior. In particular, phase transitions are possible. As the parameters (β and parameters in W, in our case) appearing in a functional integral vary, the behavior of the functional integral may also change discontinuously or at least drastically. The computational complexity involved in evaluating the functional integral may also change correspondingly. For instance, we may want to calculate how correlation lengths change with β.

Thus, in the actual experiment, it may be interesting to see to what extent the actual performance tracks the optimal performance. It may happen that for a region of the parameter space, the actual performance would agree with the optimal performance, but as the experimenter varies the parameters, the actual performance may abruptly deviate from the optimal performance, or it may drop drastically even as it tracks the optimal performance.

LIKELY AND UNLIKELY DISTORTIONS

A potential point of confusion is that while we disavow, in the context of this discussion, any theorizing about how the visual system works, the functional integral *can* be regarded as such a theory: we summon from memory storage either prototype pattern ϕ_0 or ϕ_1, apply distortion, add noise, and try to find particular forms of the distortion and noise so as to match the seen pattern ϕ, all the while weighing the likelihood of the particular distortion and noise. This "theory," while simple, is not implausible. Subjectively, we feel that when varying a pattern, we gauge the likelihood of various distortions. In other words, we carry in our heads a functional W. We can easily imagine designing a machine along these lines.

A quite different theory would suggest that we look for and identify "features" such as edges between predominantly black areas and predominantly white areas in the patterns ϕ_0 and ϕ_1.

Suppose experiments show the actual performance to be substantially below optimal performance. What would that mean? It might mean that the visual system is capable of only a crude approximation in evaluating the functional integral involved. It would then be interesting to determine what approximation the visual system uses. This is certainly possible in principle. Alternatively, it might mean that the visual system can evaluate the relevant functional integral fairly accurately, but that the W used by the experimenter does not correspond to the W that we "carry in our heads."

In principle, the experimenter can carry out a series of experiments, each with a different W, all corresponding to "reasonable" choices. Suppose the optimal performance can be determined for each W. It could happen that the actual performance does not come close to the optimal performance for any of these Ws. Perhaps more interestingly, it could also happen that the actual performance reaches or comes close to the optimal performance for some Ws.

The correspondence with statistical mechanics also suggests the question of whether some sort of universality might play an essential role in visual perception. We can also ask whether the corresponding statistical mechanical system exhibits short-ranged or long-ranged correlation. In this connection, we may perhaps emphasize that two logically distinct issues surface in our program. First, we have the question of whether the visual system can attain the optimal performance theoretically attainable. Next, given that this optimal performance is in fact attained, we can ask what computations are necessary to attain this performance.

We would like to conjecture that actual performance does in fact come close to optimal performance for some reasonable W. If experiments verify our conjecture, then we are confronted by the interesting issue of the type of circuitry and algorithm capable of effectively evaluating the functional integral involved.

LOCAL VERSUS NON-LOCAL COMPUTATIONS

In our work, we construct tasks in which arbitrarily long-ranged and multi-point correlations must be computed if optimal performance is to be reached, at least in certain limits which are controllable as the different image ensembles are generated. It is known that in discrimination among simpler image ensembles, human observers can approach optimal performance in the sense defined here (Barlow 1980, 1982). This suggests experiments in which the performance of humans is measured as a function of the parameters which control the relevant correlation lengths. If the visual system can only compute local functionals, as with feature detectors, performance should follow the optimum only for a restricted range of correlation lengths and then fall away dramatically. If, on the other hand, the system can adapt to compute strongly non-local functionals of image intensity, no such abrupt drop will be observed. These experiments will be difficult, but they have the potential of providing serious challenges to our understanding of computation in the nervous system.

Our suspicion is that the system can solve non-local problems, and that there are interesting theoretical questions to be answered about the algorithms and hardware responsible for such contributions. Suspicions aside, the approach described here (Bialek and Zee 1987, 1990b) provides the tools for asking very definite questions about the computational abilities of the brain.

Unfortunately, it is well-nigh impossible to evaluate functional integrals exactly. After all, the exact evaluation of a functional integral amounts to the exact solution of a statistical mechanical system or of a quantum field theory. The history of statistical mechanics and quantum field theory testifies amply to the difficulty of the task. Thus, in our work we are reduced to trying various approximations to the functional integrals, often reaching only qualitative conclusions. (Of course, the functional integrals can also be evaluated numerically.)

~ 487 ~

Instead of trying to evaluate $P(\phi \mid \phi_0)$, we have also tried to extract some general features. In particular, we have considered using the renormalization group to study the properties of $P(\phi \mid \phi_0)$ in an attempt to discover a strategy for "universal computation" in processing visual information (Bialek and Zee 1991b).

It is the interplay between noise and distortion that makes the evaluation of the field theory defined by $P(\phi \mid \phi_0)$ so difficult. If either noise or distortion is omitted, the task of evaluating $P(\phi \mid \phi_0)$ becomes considerably simpler. (In particular, with no distortion, the problem becomes Gaussian and trivial.) Why do we make life miserable for ourselves? Because we want to appreciate the difficulty of a task that the visual system performs extremely well (at least according to the subjective evidence). Indeed, our work represents largely a record of our awakening to how difficult the computations involved are. The difficulty of this task is also reflected in the fact that machines with artificial vision have not mastered this task of "invariant perception." Indeed, as far as we know, current machines have difficulty recognizing images if the image can be arbitrarily rigidly translated, rotated, and dilated. Of course, it may also turn out that our visual system does not perform as well as we think it does.

Models

FEATURE DETECTORS

In the second part of our work, we attempt to capture, in quantitative models, the essence of some leading theories of perception. We then compute the predicted performance at a "naturalistic" perceptual task, with the aim of ultimately comparing whatever results we may obtain with actual experiments on the human visual system.

For example, consider the feature-detector theory which originated in the neuro-physiological experiments of the 1950s.[2] Neurons in the visual system are assumed to compute nonlinear functionals of the image intensity and thus signal the presence of features in the image. Thus, the continuous pattern $\phi(x)$ is converted into a set of discrete "feature tokens" to be processed by subsequent layers of neurons. We attempt to capture the essence of this theory by taking the simplest possibility for the feature tokens: they are Ising spins σ_μ located at x_μ, $\mu = 1, 2, \ldots, N$, with σ_μ taking on values ± 1. The image is sampled at x_μ to give $\phi_\mu = \int d^2x\, f(x - x_\mu)\phi(x)$ where $f(x - x_\mu)$ represents the response function of a feature-detector neuron located at x_μ. The response function $f(x)$, with its excitatory center and inhibitory surround, is well known to neurophysiologists (Levine and Shefner 1981). It is often modeled as the Laplacian of a Gaussian, $\nabla^2 G$, or as the difference of two Gaussians.[3]

Our model (Bialek and Zee 1988) is that σ_μ tends to be $+1$ when ϕ_μ is positive and -1 when ϕ_μ is negative, as

---

[2] For a brief review of the history of feature detectors, see Barlow (1980, 1982).

[3] For example, see Kuffler (1953) and Parker and Hawken (1985); also Albrecht (1982, 117ff.).

described by some probability distribution $P(\sigma \,|\, \phi)$. Putting it together, we have the conditional probability $P(\sigma \,|\, \phi_0) = \int \mathcal{D}\phi \, P(\sigma \,|\, \phi) P(\phi \,|\, \phi_0)$. In other words, the experimenter (or the natural environment we live in) turns the known image ϕ_0 into ϕ. The seen image ϕ is then processed into the "feature tokens" σ_μ.

We believe that this "Ising" model is prototypical of a large family of models which replace the continuous image $\phi(x)$ by discrete and local feature tokens. It contains one of the classic feature-detector ideas concerning the extraction of edges, a concept formalized by Marr, Poggio, and others as the location of contours where some appropriately filtered version of the image vanishes. Here the "domain walls" between spin-up and -down regions mark the zero-crossing contours, so in fact this spin representation has a bit more information than a "sketch" based on zero-crossing contours alone.

The point is that these models are sufficiently well defined so that they can be studied numerically or analytically in certain limits. For instance, if the range of the response function $f(x)$ is small compared to the intercellular spacing (which is not biologically reasonable), the maximum efficiency can be seen to be $2/\pi = 0.64$, which appears low compared to experimental reports of efficiency ranging from 0.5 to 0.95. We conclude that overlaps of the receptive fields of neighboring cells are essential for understanding the observed efficiency of visual perception. Furthermore, these overlaps must be negative to enhance $(d')^2$, which necessitates an excitatory-center, inhibitory-surround type of organization found for real neurons. Obviously, we can go on to consider variations of this model. For instance, the work of Hübel and Wiesel established that certain cells are selectively sensitive to directions (Levine and Shefner 1981). Thus, instead of Ising spins, we can consider "Heisenberg" spins $\vec{s}_\mu$.

~489~

LINEAR FILTERS

Another class of models that we have considered supposes that the detectors in the visual system act as linear filters (Bialek and Zee 1991a). In other words, each detector functions as a narrow-band Fourier analyzer centered at some characteristic spatial frequency. Models of this type are suggested by the work of Campbell, Robson, Lawden, and De Valois.[4]

PERCEPTION BY STEEPEST DESCENT

In recent years, neural networks and related systems have been studied intensively as models of the brain. The essence of the approach consists of casting various cognitive, learning, and perceptual tasks as optimization problems. Neural networks have been shown to be able to solve optimization problems efficiently and in parallel. Thus, it is an attractive theory that the functional integrals relevant to visual perception are evaluated in the steepest descent approximation. We feel that this is an important and urgent issue that should be settled by experiment (Bialek and Zee 1990a). Does the brain merely do steepest descent? Or is it considerably more sophisticated?

To answer this question, we need to have a version of the problem outlined in the section on optimal performance, but simplified to such a degree that we can solve it both exactly and in the steepest descent approximation. We settle on a simple experiment described by

$$P(\phi \,|\, \phi_0) = \frac{1}{Z'} \int dp \, e^{-\beta \int d^2x \, [\phi(x) - \gamma \phi_0(x_p)]^2},$$

where p parameterizes a family of distortions. For example, p can be an angle θ and x_p equal to x rotated through θ. We also take $\phi_1 = 0$ for simplicity. Thus, the subject is to decide on the presence

---

[4]See DeValois (1982), Campbell and Lawden (1982), and references therein.

or absence of the prototype pattern ϕ_0 with the pattern obscured by noise and presented with a randomly chosen orientation on each trial. We are interested in the regime in which the noise becomes overwhelming. (By simple scaling, we see that quantities such as $(d')^2$ depend only on $\beta^{\frac{1}{2}}\gamma$, so the β can be absorbed.) If $\int d^2x\,\phi_0^2(x_p) = \int d^2x\,\phi_0^2(x)$, as is the case for the examples we have considered, the relevant functional integral can be organized in the suggestive form

~491~

$$P(\phi \mid \phi_0) = \frac{1}{Z} \int dp\, e^{-\beta H_0(\phi) - \gamma H_1(\phi, p)}$$

with the bare Hamiltonian $H_0 = \int d^2x\,\phi^2$ and the "perturbing" Hamiltonian $H_1 = -\beta \int d^2x\,\phi(x)\phi_0(x_p)$. Our task is to evaluate the integral over p exactly and in the steepest descent approximation and, hence, to obtain the efficiency $\epsilon = (d')^2_{\text{steepest descent}}/(d')^2$. Of course, in the large-γ limit, ϵ tends to 1 as a mathematical statement of the efficacy of the steepest descent approximation. We are interested in the opposite noisy limit. Thus, we evaluate the small-γ expansion exactly and by steepest descent.

We find to lowest order in γ that

$$(d')^2_{\text{steepest descent}} = \gamma^2 \frac{[\langle \bar{H}_1 H_1^* \rangle_0 - \langle H_1^* \rangle_0 \langle \bar{H}_1 \rangle_0]^2}{[\langle H_1^{*2} \rangle_0 - \langle H_1^* \rangle_0^2]}.$$

Here $\bar{H}_1 = \int dp\, H_1(\phi, p)$ (with the integral over p normalized so that $\int dp = \int [d\theta/2\pi]$ if p is an angle) and $H_1^* \equiv H_1(\phi, p^*(\phi))$ is the minimum of $H_1(\phi, p)$ as a function of p. Notice that it has a highly non-trivial dependence on ϕ. As indicated, the various expectation values are to be taken with the bare Hamiltonian $\langle \cdots \rangle_0 = [1/Z_0] \int \mathcal{D}\phi\, e^{-\beta H_0(\phi)}(\cdots)$. In comparison, we have $(d')^2 = \gamma^2[\langle \bar{H}_1^2 \rangle - \langle \bar{H}_1 \rangle^2]$. The obvious condition that $\epsilon \leq 1$ is satisfied by Schwarz's inequality.

The evaluation of expectation values such as $\langle \bar{H}_1 H_1^* \rangle$ is a rather non-trivial exercise in functional integration. Again, we have to use generalizations of Rice's method determining the distribution of minima. Here we merely summarize the result, which turns out to depend on the correlation function

$$\Delta(p,p') = \langle H_1(\phi,p)H_1(\phi,p') \rangle = 2\beta \int d^2x \; \phi_0(x_p)\phi_0(x_{p'}).$$

(In the examples we considered, $\Delta(p,p') = \Delta(p-p')$ is "translation invariant.") Let $\Delta_0 = \Delta(0), \Delta_2 = [d^2\Delta(0)/dp^2]$, and $\Delta_4 = [d^4\Delta(0)/dp^4]$. Then

$$(d')^2_{\text{steepest descent}} = \gamma^2 \left(\int dp\, \Delta(p) \right)^2 \frac{\Delta_4}{(\Delta_0\Delta_4 + (1-\frac{\pi}{2})\Delta_2^2)}.$$

To see what is actually going on, we can now try out various specific prototype pictures $\phi_0(x)$. For example, we have considered a wedge- or leaf-shaped picture $\phi_0(x) = f(r)e^{-[1/2C(r)]\theta^2}$ where r and θ are the polar coordinates of x. The important quantity here is the width of the wedge $C^{\frac{1}{2}}(r)$ (which we take to be small). We find

$$\epsilon = \frac{\langle C^{\frac{1}{2}} \rangle}{\pi^{\frac{1}{2}}} \frac{1}{1 + \frac{1}{3}(1 - \frac{\pi}{2})\frac{\langle C^{-1} \rangle^2}{\langle C^{-2} \rangle}}$$

where $\langle \cdots \rangle$ denotes some average of $(\cdots)$ over the radial direction weighted by $f^2(r)$ and geometrical factors. If $C(r) =$ constant (so that the picture is wedge-shaped), we have $\epsilon = [C^{\frac{1}{2}}/\pi^{\frac{1}{2}}][3/4(1 - [\pi/8])]$. If the picture is very jagged so that $\langle C^{-2} \rangle \gg \langle C^{-1} \rangle^2$, then $\epsilon = [\langle C^{\frac{1}{2}} \rangle/\pi^{\frac{1}{2}}]$.

How can we use this analysis to find out if the visual system is actually an efficient device that locates minima or "best matches" (as simple neural network models would suggest)? Suppose the experiment outlined is done and the measured efficiency comes out to be equal to the value predicted by steepest descent. That would offer dramatic support for the idea of "best match." On the other hand, if the efficiency is measured to be greater than the predicted efficiency, that would rule out or at least cast grave doubt on the "best match" theory. Unfortunately, the situation is complicated by the possibility that information is lost by processing, for instance, by feature detectors. Thus, one would have to consider the steepest descent approximation in evaluating the integral over p not in $P(\phi \,|\, \phi_0)$, but in $P(\sigma \,|\, \phi_0) = \int \mathcal{D}\phi \, P(\sigma \,|\, \phi) \, P(\phi \,|\, \phi_0)$ (with σ denoting some feature "tokens"). For the calculation outlined here to be relevant, we have to suppose that processing affects both $(d')^2_{\text{steepest descent}}$ and $(d')^2$ in the same proportion so that the effect cancels out in ϵ. Note, however, that the experiment can be repeated and the theoretical expression for ϵ can be evaluated (numerically at least) for a wide variety of prototype pictures ϕ_0.

~493~

Summary

In summary, we have outlined a systematic program[s] to address some quantitative issues that we must resolve in order to understand visual perception. An important point is that images used in vision experiments should be generated from statistical ensembles that may be formulated analytically. Advances in high-speed computation should make possible this type of controlled experiment and at the same time facilitate the analysis of models of how the human perceptual system tries to determine these statistical ensembles.

SOME QUESTIONS

In conclusion, I would like to pose the following list of questions as a challenge to vision researchers.

1. What is the optimal performance allowable for various perceptual tasks?

---

[s]A somewhat fuller account of the discussion presented here may be found in Zee (1989).

2. What are the computations needed to reach this optimal performance? Can we identify the issues involved (for instance, local versus non-local computation)?

3. Is the visual system actually capable of this optimal performance? How close does it come? (These questions can be answered only by experiments, of course.)

4. If the performance of the visual system approximates optimal performance, how does it perform the computations identified in question 2 above? What neural circuitry and algorithm can carry out these computations? Can various simple models be ruled out?

5. Does the visual system operate by optimization? Is the performance reached by the steepest descent approximation in accordance with observation?

6. Are there universal features and properties in the sense of statistical physics?

As is evident from the preceding discussion, we have touched only on the beginnings of this program and have reached only qualitative conclusions at best. Many challenging problems remain. ✦

Acknowledgments

I am indebted to W. Bialek for numerous stimulating and interesting discussions. This research was supported in part by the National Science Foundation under grant no. PHY82-17853, supplemented by funds from the National Aeronautics and Space Administration, at the University of California at Santa Barbara.

REFERENCES

Albrecht, D. G., ed. 1982. *Recognition of Pattern and Form*. New York, NY: Springer-Verlag. doi:10.1007/978-3-642-93199-4.

Barlow, H. B. 1980. "The Absolute Efficiency of Perceptual Decisions." *Philosophical Transactions of the Royal Society B* 290 (1038): 71–82. doi:10.1098/rstb.1980.0083.

———. 1982. "The Past, Present, and Future of Feature Detectors." In *Recognition of Pattern and Form*, edited by D. G. Albrecht, 4–32. New York, NY: Springer-Verlag. doi:10.1007/978-3-642-93199-4.

Bialek, W., and A. Zee. 1987. "Statistical Mechanics and Invariant Perception." *Physical Review Letters* 58 (7): 741–44. doi:10.1103/physrevlett.58.741.

———. 1988. "Understanding the Efficiency of Human Perception." *Physical Review Letters* 61 (13): 1512–15. doi:10.1103/physrevlett.61.1512.

———. 1990a. "Inadequacy of Mean Field Approximation in Visual Perception." In preparation.

———. 1990b. "Invariant Perception: A Functional Integral and Field Theoretic Approach." In preparation.

———. 1991a. "Linear Filter Models in Visual Perception." In preparation.

———. 1991b. "Recognizing Ensembles of Images: Universality at Low Resolution." In preparation.

Campbell, F. W., and M. Lawden. 1982. "The Physics of Visual Perception." In *Recognition of Pattern and Form*, edited by D. G. Albrecht, 146–51. New York, NY: Springer-Verlag. doi:10.1007/978-3-642-93199-4.

DeValois, R. L. 1982. "Early Visual Processing: Feature Detection or Spatial Filtering." In *Recognition of Pattern and Form*, edited by D. G. Albrecht, 152–74. New York, NY: Springer-Verlag. doi:10.1007/978-3-642-93199-4.

Kuffler, S. W. 1953. "Discharge Patterns and Functional Organization of Mammalian Retina." *Journal of Neurophysiology* 16 (1): 37–68. doi:10.1152/jn.1953.16.1.37.

Levine, M. W., and J. M. Shefner. 1981. *Fundamentals of Sensation and Perception*. New York, NY: Random House.

Marr, D. 1982. *Vision*. New York, NY: W. H. Freeman.

Parker, A., and M. Hawken. 1985. "Capabilities of Monkey Cortical Cells in Spatial-Resolution Tasks." *Journal of the Optical Society of America A* 2 (7): 1101–14. doi:10.1364/josaa.2.001101.

Zee, A. 1989. "Some Quantitative Issues in the Theory of Perception." In *Evolution, Learning and Cognition,* edited by Y. C. Lee, 183–216. Singapore: World Scientific. doi:10.1142/9789814434102_0007.

PROBABILITY,
ENTROPY & QUANTUM

66

THERMODYNAMIC CONSTRAINTS ON QUANTUM AXIOMS

Asher Peres, Israel Institute of Technology

The second law of thermodynamics imposes severe constraints on the formal structure of quantum theory. In particular, that law would be violated if it were possible to distinguish non-orthogonal states or if Schrödinger's equation were nonlinear.

Introduction

Thermodynamics, relativity, and quantum theory are the three pillars upon which the entire structure of theoretical physics is built. They are not branches of physics (like acoustics, optics, etc.) but general frameworks encompassing every aspect of physics. Thermodynamics—for which a more appropriate name would have been "thermostatics"—governs the convertibility of various forms of energy; relativity theory deals with measurements of space and time; and quantum theory is a set of rules for computing probabilities of outcomes of tests (also called "measurements") following specified preparations (Stapp 1972).

Each member of this triad involves *time-ordering* as a primitive concept. In thermodynamics, high-grade ordered energy can spontaneously degrade into a disordered form of energy, called *heat.* The time-reversed process never occurs. More technically, the total entropy of a closed physical system cannot decrease. In relativity, information is collected from the *past* light cone and propagates into the *future* light cone.

And in quantum theory, probabilities can be computed for the outcomes of tests which *follow* specified preparations, not those which *precede* them.

Specific physical situations may involve two of these fundamental frameworks, or even all three simultaneously. High-energy astrophysics is an obvious example. Another relatively simple problem is the equilibrium of electromagnetic radiation with a relativistic gas in an isothermal enclosure. These situations raise the question of the mutual consistency of seemingly unrelated requirements imposed by thermodynamics, relativity, and quantum theory. For example, a detailed investigation of the radiating relativistic gas, along the lines of the classic work of Einstein (1917), shows that thermal equilibrium can be obtained if, and only if, the spontaneous decay rate of an excited atom (Einstein's *A*-coefficient) is reduced in the exact ratio of the relativistic time dilation due to the motion of that atom (Peres 1981).

The purpose of this chapter is to show that tampering with some of the axioms of quantum theory may lead to a violation of the second law of thermodynamics. For example, this would happen if there were a method for distinguishing nonorthogonal states or if Schrödinger's equation were nonlinear. The reason for this violation is that *entropy*, as defined in quantum theory by von Neumann (1955) and in information theory by Shannon (1948), is fully equivalent to the mundane entropy of mechanical engineers. In particular, if the von Neumann–Shannon entropy of a closed physical system can decrease (as a result of a modification of the axioms of quantum theory), it becomes possible to build conceptual engines extracting an unlimited amount of work from an isothermal reservoir.

In the next section, I shall prove (that is, I shall argue as convincingly as I can) that the von Neumann–Shannon entropy *is* authentic entropy, with all the rights and privileges pertaining thereto. That section is the difficult part of this chapter. The reader who is already convinced that the claim is true may skip the long "proof." Various corollaries are derived in the following section.

Entropy

The purpose of this section is to prove that von Neumann's definition of entropy is equivalent to that of standard thermodynamics. I hope that this proof will be more readable, and also more convincing, than the one found in the classic book by von Neumann (1955). I shall use for this proof some recent results due to Partovi (1989).

The entropy of a mixture of dilute, inert, ideal gases is a standard problem of classical thermodynamics (Zemansky 1968). Its value is

$$S = -N \sum_j c_j \log c_j, \qquad (1)$$

where N is the total number of molecules and c_j is the concentration of the jth species. (Units are chosen so that Boltzmann's constant equals 1. Temperature is therefore measured in energy units and entropy is dimensionless.) The derivation of equation 1 relies on the possibility of making semipermeable membranes which are transparent to type-j molecules and opaque to all others. These membranes are used as pistons in an ideal frictionless engine, immersed in an isothermal bath at temperature T, as sketched in figure 1. It is easily shown (Zemansky 1968) that a reversible separation of the mixed gases must supply an amount of isothermal work equal to $-NT \sum c_j \log c_j$. This work is converted into heat and released into the reservoir. Therefore the mixing entropy is given by equation 1.

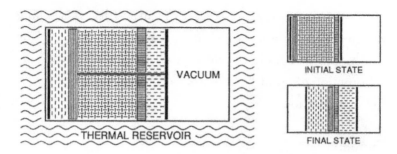

Figure 1. Ideal engine used to separate gases A (to the left) and B (to the right). The vertically and horizontally hatched semipermeable pistons are transparent to gases A and B, respectively. The mechanical work that must be supplied in order to transform the initial state into the final state is released as heat into the thermal bath.

Von Neumann's definition of entropy of a quantum state closely parallels the above argument. It assumes that there are semipermeable membranes capable of separating orthogonal states with 100% efficiency—this is indeed the operational meaning of "orthogonal states." The fundamental problem is whether it is legitimate to treat quantum *states* in the same way as classical ideal gases, and in particular why one should expect thermal equilibrium to be achieved.

In his proof, von Neumann (1955) relies on a subterfuge proposed by Einstein (1914) in the early days of the "old" quantum theory. Consider many similarly prepared quantum systems, such as Bohr's planetary atoms. Each one is enclosed in a large box with impenetrable walls, so as to prevent any interaction between these quantum systems. All these boxes are then placed into an even larger container, where they behave as an ideal gas, because each box is so massive that classical mechanics is valid for its motion (i.e., there is no need of Bohr–Sommerfeld quantization rules—remember that we are in 1914). The container itself has ideal walls which may be, according to our needs, perfectly conducting, perfectly insulating, or with properties equivalent to those of semipermeable membranes.

These "membranes" are endowed with automatic devices able to peek inside the boxes and to test the state of the quantum systems enclosed therein. In his book, von Neumann (1955, 359) insists that the practical infeasibility of this contraption does not impair its demonstrative power: "In the sense of phenomenological thermodynamics, each conceivable process constitutes valid evidence, provided that it does not conflict with the two fundamental laws of thermodynamics." He then shows that equation 1 can be recast in the form

$$S = -N \operatorname{Tr}(\rho \log \rho), \qquad (2)$$

where ρ is the *density matrix* representing the state of a molecule of our gas. The c_j of equation 1 correspond to the eigenvalues of ρ.

The problem remains as to whether this hybrid classical–quantal reasoning is consistent. My purpose here is to give a genuinely quantal proof of the equivalence of von Neumann's entropy, equation 2, to the ordinary entropy of classical thermodynamics.

A SUMMARY OF QUANTUM THEORY

As stated above, the essence of quantum theory is that it allows one to compute *probabilities* for the outcomes of tests, following specified preparations (Stapp 1972). It therefore is a *statistical* theory: When we want to make predictions about a specific quantum object, such as an atom, we must consider it as a member of a *Gibbs ensemble*, namely an infinite set of conceptual replicas of that object, all identically prepared. Only then can we give a meaning to the notion of probability. The fundamental assumption of quantum theory is that *all* information about the preparation of such an ensemble can be represented by a Hermitian matrix ρ satisfying $\operatorname{Tr} \rho = 1$, called the density matrix.

Quantum theory further assumes that every observable property A of a physical system is also represented by a Hermitian matrix. I shall denote this matrix by the same letter A, for simplicity. If property A is actually measured on a quantum system, then, for *any* preparation of that system, the result of the measurement turns out to be one of the eigenvalues of the matrix A. Moreover, if we have numerous quantum systems, each one resulting from the same preparation represented by the matrix ρ, and if A is measured on each one of them individually, the average value of the results tends to $\langle A \rangle = \text{Tr}(A\rho)$. The latter formula is derived by von Neumann from fairly weak assumptions (see von Neumann 1955, 316).

These rules have a remarkable consequence. Given two different preparations represented by matrices ρ_1 and ρ_2, one can prescribe another preparation ρ by the following recipe: Let a random process have probability λ of "success" and probability $1 - \lambda$ of "failure." In the case of success, prepare the quantum system according to ρ_1. In the case of failure, prepare it according to ρ_2. This process results in a ρ given by

$$\rho = \lambda\rho_1 + (1 - \lambda)\rho_2. \tag{3}$$

Indeed, if the above instructions are executed a large number of times, the *average* value obtained for subsequent measurements of A is

$$\langle A \rangle = \lambda \, \text{Tr}(A\rho_1) + (1 - \lambda) \, \text{Tr}(A\rho_2) = \text{Tr}(A\rho). \tag{4}$$

What I find truly amazing in this result is that once ρ is given, it contains *all* the available information, and it is impossible to reconstruct from it ρ_1 and ρ_2! For example, if we prepare a large number of polarized photons and if we toss a coin to decide, with equal probability, whether the next photon to be prepared will have vertical or horizontal *linear* polarization, or, in a *different experimental set-up*, we likewise randomly decide whether each photon will have right-handed or left-handed *circular* polarization, we get in both cases the same

$$\rho = \frac{1}{2}\begin{pmatrix} 1 & 0 \\ 0 & 1 \end{pmatrix}.$$

An observer receiving megajoules of these photons will never be able to discover which one of these two methods was chosen for their preparation, notwithstanding the fact that these preparations are *macroscopically* different. (If this were not true, Einstein–Podolsky–Rosen [EPR] correlations would allow instantaneous transfer of information to distant observers, in violation of relativistic causality; see Herbert [1982].)

Another example would be to prepare photons having, with equal probability, linear vertical polarization or circular right-handed polarization. An observer requested to guess what the preparation of a particular photon was, under the best conditions allowed by quantum theory, would be able to give the answer *with certainty* in only 29.3% of cases (Ivanovic 1987; Peres 1988). It will be shown below that a "superobserver" who could always give an unambiguous answer would also be able to extract an infinite amount of work from an isothermal reservoir.

THE ENTROPY OF A QUANTUM ENSEMBLE

Our first task is to derive von Neumann's formula in equation (1) without invoking Einstein's classical impenetrable boxes (Einstein 1914). In order to do so, we shall replace these fictitious boxes by *quantum* degrees of freedom—which may be equally fictitious, but at least cannot lead to any of the inconsistencies that plague hybrid quantal–classical systems.

Let q denote collectively all the (real) degrees of freedom and let $\mathbf{R}$ be three additional, fictitious degrees of freedom of our quantum system (they can be interpreted as the center-of-mass coordinates of the Einstein's box enclosing it). The wave function $\psi(q)$ therefore becomes $\psi(q, \mathbf{R})$—there is no danger of inconsistency here, since $\mathbf{R}$ belongs to another, independent *quantum* system. We then write for the Hamiltonian

$$H = H_0 + \frac{\mathbf{P}^2}{2M},\qquad(5)$$

where H_0 is the original Hamiltonian of the quantum system, $\mathbf{P}$ is the momentum conjugate to $\mathbf{R}$, and M is an arbitrarily large number. At this stage, there is no interaction between the original degrees of freedom and the new fictitious ones. If we want to use density matrices rather than wave functions, we write these as $\rho(q'\mathbf{R}', q''\mathbf{R}'')$.

Next, we introduce Einstein's large container, which also serves as a thermal reservoir. It may have moving parts, such as pistons, to suit our needs. It is convenient to divide its degrees of freedom into two classes: a small number of macrovariables (center-of-mass position, spatial orientation, location of the pistons, etc.), collectively denoted by X, and a huge number of microvariables, denoted by x, which describe the atomic structure of the container. The macrovariables have a very slow motion and therefore can be treated classically in the Born–Oppenheimer approximation. The microvariables are in thermal equilibrium. Their density matrix is a Gibbs state at temperature β^{-1}:

$$\rho_1 = \frac{\exp(-\beta H_1)}{Z},\qquad(6)$$

where H_1 is the Hamiltonian of the container (in the Born–Oppenheimer approximation) and $Z = \mathrm{Tr}[\exp(-\beta H_1)]$. We further assume that the energy uncertainty of the quantum system enclosed in the container far exceeds the average level spacing of the container energy spectrum, so that the quantum system does not feel this spectrum discreteness (Chirikov 1986).

Recall that the extra degrees of freedom **R** of the quantum system exist only in our imagination, just as the thermal reservoir with its moving pistons. Nevertheless, as long as their existence does not violate known laws of physics, their introduction is a perfectly legitimate method to discover additional laws. (Their role is analogous to that of the fictitious "observers" who send and receive signals along the light cones of relativity theory.)

Finally, we introduce an interaction between our quantum system and the container in which it is enclosed. The result of that interaction can be considered as a *scattering* of the quantum system (with dynamical variables q, **R**, and **P**) due to its collisions with the container. Recall that the latter is described by microvariables x (quantized) and macrovariables X (considered now as classical parameters).

Two cases must be distinguished, depending on whether the container includes semipermeable partitions or only walls which are indifferent to the internal state of the quantum system. In the latter case, the interaction Hamiltonian involves x, X, and **R** (and possibly **P**) but not the original variables q of the quantum system. The internal evolution of the latter thus is completely disjoint from that of the fictitious variables x, X, **R**, and **P**. On the other hand, a semipermeable piston selecting quantum systems according to their internal state *must* be described by an interaction term which also involves q.

We can now apply a theorem recently derived by Partovi (1989). If before the scattering event the container is in the Gibbs state given by equation (6), the property of the scattered quantum system represented by the expression

$$S - \beta(E) = -\mathrm{Tr}(\rho \log \rho) - \beta \, \mathrm{Tr}(\rho H_0) \qquad (7)$$

cannot decrease as a result of the collision. This follows from conservation of energy and convexity of entropy (Wehrl 1978). Note that in equation (7), ρ and H_0 refer to the quantum system, but β refers to the thermal reservoir with which it collided.

We further assume that the thermal reservoir is so large that its state after the collision can again be described by equation (6), with the same β, and in particular that it is justified to ignore its correlation with the state of the quantum system. We therefore expect that after numerous collisions, $S - \beta\langle E \rangle$ will reach the maximum value allowed by selection rules. If all the states of the quantum system are accessible (i.e., if there are no selection rules), the maximum value of $S - \beta\langle E \rangle$ corresponds to a Gibbs state of the quantum system, at the same temperature β^{-1} as the reservoir. Not every state, however, may be accessible. In particular, if the container has passive walls interacting only with the **R** degrees of freedom but not with the q variables, the internal state of the quantum system cannot be affected. In that case, an ensemble of quantum systems described by equation (5) has the same statistical properties as a classical ideal gas of free particles of mass M. In particular, it exerts exactly the same pressure on the walls of the container. This is an immediate consequence of the evolution equation of the Wigner distribution function (Wigner 1959), which, for free particles, is identical to the Liouville equation in classical statistical mechanics.

Up to this point, nothing was been proved that has any consequence in the real world. Only the fictitious degrees of freedom **R** were thermalized by multiple collisions with the fictitious container. The situation becomes more interesting if semipermeable partitions are introduced. As explained above, these partitions are described by an interaction term involving q, **R**, and X (note that the classical parameters X are prescribed functions of time and that their time dependence must be extremely slow on time scales relevant to the quantum system; otherwise the Born–Oppenheimer approximation would not be valid).

To describe a semipermeable partition, we have to add to the right-hand side of equation 5 a term which, in the simplest case, has the form $V(q, \mathbf{R}, X)$. This term, if suitably chosen, causes the formation of correlations between the variables q and $\mathbf{R}$. For example, we can concentrate particles with spin up in one part of the container and those with spin down in the other part by introducing an interaction

$$H_{\text{int}} = -\sigma_2 \, \mathbf{R} \cdot \mathbf{n} \, F(\mathbf{R}, X), \qquad (8)$$

where $\mathbf{n}$ is a constant unit vector and F is a non-negative function of its arguments (in particular, F may be large only when $\mathbf{R}$ approaches some combination of X representing the presence of a semipermeable partition). The interaction described by equation 8 corresponds to a force $\mathbf{n}F\sigma_z$ acting on the quantum system. Therefore $\mathbf{R}$ will accelerate in the direction $\pm\mathbf{n}$, the $\pm$ sign being that of the eigenvalue of σ_z corresponding to the state of the quantum system.

This interaction is readily generalized to one that sorts different eigenvalues $a_1, a_2, \ldots$ of any operator A. Writing A instead of σ_2 in equation 8, we find that the quantum systems are subject to forces $\mathbf{n}F a_1$, $\mathbf{n}F a_2$, etc., according to their state. The lesson learned from this simple theoretical model is that nothing prevents us, at least in principle, from confining in different regions of $\mathbf{R}$-space quantum systems in *orthogonal* states. They behave exactly as if they were a mixture of classical ideal gases. Therefore, there should be no doubt that the von Neumann entropy is *equivalent* to the entropy of classical thermodynamics (in the same sense that the quantum notions of energy, momentum, angular momentum, etc. are equivalent to the classical notions bearing the same names).

How to Beat the Second Law of Thermodynamics

SELECTING NON-ORTHOGONAL STATES

It is noteworthy that the interaction described by equation (8) and its generalizations allow us to distinguish *different eigenvalues* of Hermitian operators, which correspond to *orthogonal* states of the quantum system. Let us suppose now that some other type of interaction would allow us to distinguish non-orthogonal states. This would have momentous consequences; for example, a certain type of EPR correlation could be used to transfer information instantaneously (Datta and Home 1986). I shall now show that this could also be used to convert into work an unlimited amount of heat extracted from an isothermal reservoir.

The process is illustrated in figure 2 for the case of two non-orthogonal photon states. Suppose that there are n photons. One half of them are prepared with vertical linear polarization and the other half with a linear polarization at 45° from the vertical. Initially, as shown in (a), they occupy two chambers with equal volumes. The first step of the cyclic process is an isothermal expansion doubling these volumes, as shown in (b). This expansion supplies an amount of work $nT \log 2$, where T is the temperature of the reservoir (recall that Boltzmann's constant k equals 1 here). At that stage, the impenetrable partitions separating the two photon gases are replaced by semipermeable membranes, similar to those of figure 1. These membranes, however, have the unusual ability of selecting *non-orthogonal* states: one of them is transparent to vertically polarized photons and reflects those with polarization at 45° from the vertical; the other membrane has the opposite properties. A double frictionless piston, like the one in figure 1, thus brings the engine to state (c) in figure 2 without expenditure of work or heat transfer. We thereby obtain a *mixture* of the two polarization states. Its density matrix is

$$\rho = \frac{1}{2} \left[\begin{pmatrix} 1 \\ 0 \end{pmatrix} \begin{pmatrix} 1 \\ 0 \end{pmatrix}^{\dagger} + \frac{1}{2} \begin{pmatrix} 1 \\ 1 \end{pmatrix} \begin{pmatrix} 1 \\ 1 \end{pmatrix}^{\dagger} \right] = \begin{pmatrix} 0.75 & 0.25 \\ 0.25 & 0.25 \end{pmatrix}. \quad (9)$$

The eigenvalues of ρ are 0.854 (corresponding to photons polarized at 22.5° from the vertical) and 0.146 (for the opposite polarization). We now replace the "unusual" membranes by ordinary ones, selecting these two orthogonal polarization states. The next step is an isothermal compression, leading to state (d) where the two chambers have the same pressure and the same total volume as those in state (a). This isothermal compression requires an expenditure of work equal to

$$- nT(0.146 \log 0.146 + 0.854 \log 0.854) = 0.416 \, nT, \quad (10)$$

which is released as heat into the reservoir. This is *less* than the amount $nT \log 2$—which was gained in the isothermal expansion from (a) to (b)—by the amount $0.277 \, nT$. Finally, no work is involved in returning from (d) to (a) by suitable rotations of polarization vectors (see von Neumann 1955, 366). We have thereby demonstrated the existence of a closed cycle whereby heat is extracted from an isothermal reservoir and converted into work, in violation of the second law of thermodynamics.

NONLINEAR SCHRÖDINGER EQUATION

A similar violation arises if nonlinear "corrections" are introduced in Schrödinger's equation, as proposed by many authors, with various motivations (de Broglie 1956; Rosen 1945; Weinberg 1989). A nonlinear Schrödinger equation does not

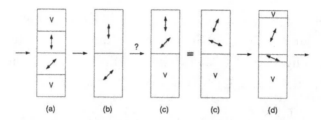

Figure 2. Cyclic process extracting heat from an isothermal reservoir and converting it into work, by using a semipermeable partition which selects non-orthogonal photon states. Double arrows represent the linear polarizations of photon ensembles. The symbol V is for vacuum.

violate the superposition principle in its weak form. The latter merely asserts that the pure states of a physical system can be represented by rays in a complex linear space. This principle does not demand that the time evolution of the rays obey a linear equation.

It is not difficult to invent nonlinear variants of Schrödinger's equation such that if $\phi(0)$ evolves into $\phi(t)$ and $\psi(0)$ evolves into $\psi(t)$, the pure state represented by $\phi(0)+\psi(0)$ does *not* evolve into $\phi(t)+\psi(t)$ but into some other pure state. I shall now show that these nonlinearities lead to a violation of the second law of thermodynamics *if the other postulates of quantum mechanics are kept intact*. In particular, I retain the fundamental assumption that "two states ϕ and ψ can be divided by a semipermeable wall if they are orthogonal" (see von Neumann 1955, 370).

Consider a *mixture* of quantum systems represented by a density matrix

$$\rho = \lambda P_\phi + (1 - \lambda)P_\psi, \tag{11}$$

where $0 < \lambda < 1$ and where P_ϕ and P_ψ are projection operators onto the pure states ϕ and ψ, respectively. The non-vanishing eigenvalues of ρ are

$$w_j = \tfrac{1}{2} \pm \left[\tfrac{1}{4} - \lambda(1 - \lambda)(1 - x)\right]^{\frac{1}{2}}, \tag{12}$$

where $x = |\langle\phi|\psi\rangle|^2$. The entropy of this mixture, $S = -k\sum w_j \log w_j$, satisfies $dS/dx < 0$ for any λ. Therefore, if the pure quantum states evolve as $\phi(0) \to \phi(t)$ and $\psi(0) \to \psi(t)$, the entropy of the mixture ρ shall not decrease (i.e., that mixture shall not become less homogeneous) provided that

$$|\langle\phi(t)|\psi(t)\rangle|^2 \leq |\langle\phi(0)|\psi(0)\rangle|^2. \tag{13}$$

In particular, if $\langle\phi(0)|\psi(0)\rangle = 0$, we must have $\langle\phi(t)|\psi(t)\rangle = 0$. Orthogonal states must remain orthogonal.

~ 513 ~

Consider now a complete orthogonal set $\{\phi_k\}$. We have, for every ψ,

$$\sum_k |\langle\phi_k|\psi\rangle|^2 \equiv 1. \tag{14}$$

Therefore, if there is some m for which $|\langle\phi_m(t)|\psi(t)\rangle|^2 < |\langle\phi_m(0)|\psi(0)\rangle|^2$, there must also be some n for which $|\langle\phi_n(t)|\psi(t)\rangle|^2 > |\langle\phi_n(0)|\psi(0)\rangle|^2$. Then the entropy of a mixture of ϕ_n and ψ will spontaneously decrease in a closed system, in violation of the second law of thermodynamics.

If we want to retain the second law, we must have $|\langle\phi(t)|\psi(t)\rangle|^2 = |\langle\phi(0)|\psi(0)\rangle|^2$ for every ϕ and ψ. It then follows from Wigner's theorem (Wigner 1932, 1959) that phases can be chosen in such a way that for every ψ, the mapping $\psi(0) \to \psi(t)$ is either unitary or antiunitary. The second alternative is ruled out by continuity. The evolution of quantum states is unitary—Schrödinger's equation must be linear—if we want to retain the other axioms of quantum theory and also the second law of thermodynamics.

Summary and Outlook

The advent of quantum theory solved one of the outstanding puzzles of classical thermodynamics, Gibbs's paradox (Landé 1953). Conversely, thermodynamics imposes severe constraints on the axioms of quantum theory. The second law of thermodynamics would be violated if it were possible to distinguish non-orthogonal states, if Schrödinger's equation were nonlinear, or if a single quantum could be cloned (Wootters and Zurek 1982). All these impossibilities are related to each other: non-orthogonal states could easily be distinguished if single quanta could be cloned, but this is forbidden by the no-cloning theorem (Wootters and Zurek 1982), which in turn follows from the linearity of Schrödinger's equation.

The key to the above claims is the equivalence of the von Neumann–Shannon entropy to the ordinary entropy of thermodynamics. The proof of this equivalence given here relies on the introduction of a mock Hilbert space with fictitious degrees of freedom. This is a perfectly legitimate way of proving theorems. However, this proof, as well as that of von Neumann, assumes the validity of Hamiltonian dynamics (in order to derive the existence of thermal equilibrium), and this point is suspicious. It may be unfair to invoke Hamiltonian dynamics in order to prove some theorems, and then to claim as a corollary that non-Hamiltonian dynamics is inconsistent. Thus, the final conclusion of the present work is that if the integrity of the axiomatic structure of quantum theory is not respected, then *every* aspect of the theory has to be reconsidered *ab initio*. ⟡

Acknowledgment

This work was supported by the Gerard Swope Fund and by the Fund for Encouragement of Research at Technion.

REFERENCES

Chirikov, B. V. 1986. "Transient Chaos in Quantum and Classical Mechanics." *Foundations of Physics* 16 (1): 39–49. doi:10.1007/bf00735179.

Datta, A., and D. Home. 1986. "Quantum Non-Separability versus Local Realism: A New Test Using the $B^0\overline{B}^0$ System." *Physics Letters A* 119 (1): 3–6. doi:10.1016/0375-9601(86)90633-x.

de Broglie, L. 1956. *Une Tentative d'Interprétation Causale et Nonlinéaire de la Mécanique Ondulatoire.* Paris, France: Gauthier-Villars.

Einstein, A. 1914. "Beiträge zur Quantentheorie." *Verh. Deut. Phys. Gesell.* 16:820–28.

———. 1917. "Quantentheorie der Strahlung." *Phys. Z* 18:121–28.

Herbert, N. 1982. "FLASH—A Superliminal Communicator Based Upon a New Kind of Quantum Measurement." *Foundations of Physics* 12:1171–79. doi:10.1007/BF00729622.

Ivanovic, I. D. 1987. "How to Differentiate between Non-Orthogonal States." *Physics Letters A* 123 (6): 257–59. doi:10.1016/0375-9601(87)90222-2.

Landé, A. 1953. *Foundations of Quantum Theory.* 10–13. New Haven, CT: Yale University Press.

Partovi, M. H. 1989. "Quantum Thermodynamics." See also contribution to the present volume, *Physics Letters A* 137 (9): 440–44. doi:10.1016/0375-9601(89)90221-1.

Peres, A. 1981. "Relativity, Quantum Theory, and Statistical Mechanics are Compatible." *Physical Review D* 23 (6): 1458–59. doi:10.1103/physrevd.23.1458.

———. 1988. "How to Differentiate between Non-Orthogonal States." *Physics Letters A* 128 (1–2): 19. doi:10.1016/0375-9601(88)91034-1.

Rosen, N. 1945. "On Waves and Particles." *Journal of the Elisha Mitchell Scientific Society* 61 (1–2): 67–73.

Shannon, C. E. 1948. "A Mathematical Theory of Communication." *Bell System Technical Journal* 27 (4): 623–56. doi:10.1002/j.1538-7305.1948.tb00917.x.

Stapp, H. P. 1972. "The Copenhagen Interpretation." *American Journal of Physics* 40 (8): 1098–116. doi:10.119/1.1986768.

von Neumann, J. 1955. *Mathematical Foundations of Quantum Mechanics.* 358–79. Princeton, NJ: Princeton Univ. Press.

Wehrl, A. 1978. "General Properties of Entropy." *Reviews of Modern Physics* 50 (2): 221–60. doi:10.1103/revmodphys.50.221.

Weinberg, S. 1989. "Particle States as Realizations (Linear or Nonlinear) of Spacetime Symmetries." *Nuclear Physics B - Proceedings Supplements* 6:67–75. doi:10.1016/0920-5632(89)90403-9.

Wigner, E. 1932. "On the Quantum Correction For Thermodynamic Equilibrium." *Physical Review* 40 (5): 749–59. doi:10.1103/physrev.40.749.

———. 1959. *Group Theory*. 233–236. New York, NY: Academic Press.

Wootters, W. K., and W. H. Zurek. 1982. "A Single Quantum Cannot Be Cloned." *Nature* 299 (5886): 802–3. doi:10.1038/299802a0.

Zemansky, M. W. 1968. *Heat and Thermodynamics*. 561–62. New York, NY: McGraw-Hill.

6 }

ENTROPY AND QUANTUM
MECHANICS

M. Hossein Partovi, California State University, Sacramento

Entropy is a natural and powerful idea for dealing with fundamental problems of quantum mechanics. Recent results on irreversibility and quantum thermodynamics, reduction and entropy increase in measurements, and the unification of uncertainty and entropy demonstrate the fact that entropy is the key to resolving some of the longstanding problems at the foundations of quantum theory and statistical mechanics.

Introduction

A distinctive feature of quantum theory is the highly non-trivial manner in which information about the quantum system is inferred from measurements. This feature is obscured in most discussions by the assumption of idealized measuring devices and pure quantum states. While for most practical purposes these are useful and reasonable approximations, it is important in dealing with fundamental issues to recognize their approximate nature. This recognition follows from the simple observation that in general measuring devices cannot fully resolve the spectrum of the physical observable being measured, a fact that is self-evident in the case of observables with continuous spectra. A consequence of this remark is that in general realizable quantum states cannot be pure and must be represented as mixed states (Blankenbecler and Partovi 1985). Equivalently, a quantum measurement in general is incomplete in that it fails to provide an exhaustive determination of the state of the system. The problem of incomplete information, already familiar from statistical mechanics, communication theory, and other areas, is thus seen to lie at the very heart of quantum theory. In this sense, quantum mechanics is a statistical theory at a very basic level, and there should be little doubt that entropy, a key idea in dealing with incomplete information, should turn out to play a central role in quantum mechanics as well. The main purpose of the following account is to demonstrate this assertion by means of examples drawn from recent work on the subject.

Entropy

To define entropy at the quantum level, we shall start with the notion of entropy associated with the measurement of an observable, the so-called *measurement entropy*. We shall then show that *ensemble entropy*, given by the well-known von Neumann formula, follows from our definition of measurement entropy by a straightforward reasoning. Later, we will establish the identity of this ensemble entropy with thermodynamic entropy, so that there will be no distinction between "information" and "physical" entropies (other than the fact that, strictly speaking, the latter is only defined for equilibrium states).

In general, a measurement/preparation process involves a measuring device D designed to measure some physical observable A. Let $\hat{\rho}$ be the density matrix representing the state of the system and $\hat{A}$ the operator representing observable A. Thus the quantum system is a member of an ensemble of similarly produced copies, some of which are subjected to interaction with the measuring device and serve to determine the state of the ensemble. The measuring device, on the other hand, involves a partitioning of the range of possible values of $\hat{A}$ into a (finite) number of *bins*, $\{\alpha_1\}$, and for each copy of the system measured it determines in which bin the system turned up. In this way, a set of probabilities $\{\mathcal{P}_i^A\}$ is determined. What becomes of the state of the copies that actually interact with the measuring device is an important question (to be discussed later), but one that is distinct from the issue of the state of the ensemble. Indeed, many measurement processes are partially or totally destructive of the measured copies of the system. The purpose of the measurement/preparation process is thus to gain information about the state of the ensemble by probing some of its members, often altering or even destroying the latter in the process.

The fact that $\hat{A}$ represents a physical observable insures that any partition of its spectrum generates a similar (orthogonal) partition of the Hilbert space, given by a complete collection of projection operators $\{\hat{\pi}_i^A\}$ in one-to-one correspondence to $\{\alpha_i\}$. Furthermore, quantum mechanics tells us that the measured probabilities are given by $\mathcal{P}_i^A = \mathrm{tr}(\hat{\pi}_i^A \hat{\rho})$, where "tr" denotes the trace operation.

How much information has the measurement produced on the possible values of A? Equivalently, how much uncertainty is there about the value of A in a given measurement? Following the work of Shannon (1948), we have a well-established information-theoretic answer to this question. This is the quantity

$$S(\hat{\rho} \,|\, D) = -\sum_i \mathcal{P}_i^A \ln \mathcal{P}_i^A, \qquad (1)$$

which will be called *measurement entropy*. It is a non-negative quantity which equals zero when one of the quantities $\mathcal{P}_i^A$ equals unity (hence no uncertainty) and equals $\ln N$ when the probabilities are all equal to $1/N$ (hence maximum uncertainty); N is the number of bins. Note that $S(\hat{\rho} \,|\, D)$ is a joint property of the system and the measuring device. Indeed, if a device with a finer partition (i.e., higher resolution) is used, the resulting measurement entropy will in general be greater. In fact it is useful to consider a *maximal device*, designated D_{max}, which is defined to be the idealized limit of a sequence of devices with ever-finer partitions. Clearly, $0 \leq S(\hat{\rho}\,|\,D) \leq \ln N$ and $S(\hat{\rho}\,|\,D) \leq S(\hat{\rho}\,|\,D_{max})$.

Consider now a series of measurements, involving observables A^ν, measuring devices D^ν, partitions $\{\alpha_i^\nu\}$, measured probabilities $\{P_i^\nu\}$, and measurement entropies $S(\hat{\rho}\,|\,D^\nu)$. Each of these entropies describes the uncertainty appropriate to the corresponding measurement. Is there an entropy that properly describes the uncertainty appropriate to the system as a whole, regardless of the individual measurements and how they were carried out? Clearly, such an overall measure of uncertainty should be gauged against devices with the highest possible resolution, i.e., against maximal devices D_{max}^ν. Moreover, if for two operators $\hat{A}$ and $\hat{B}$, $S(\hat{\rho}\,|\,D_{max}^A) < S(\hat{\rho}\,|\,D_{max}^B)$, then A must be deemed a better representative of the available information on $\hat{\rho}$ than B. From these two requirements we conclude that the quantity we are seeking is given by the least upper bound of measurement entropies $S(\hat{\rho}\,|\,D_{max}^A)$ as A is varied over all possible observables:

$$S(\hat{\rho}) = \inf_A S(\hat{\rho}\,|\,D_{max}^A). \tag{2}$$

One can show that the right-hand side of equation 2 is realized for $\hat{A} = \hat{\rho}$. The corresponding minimum is then found to be the von Neumann expression for ensemble entropy, $-\mathrm{tr}\,\hat{\rho}\ln\hat{\rho}$. Starting from the elementary definition for measurement entropy given in equation 1, we have thus arrived at the standard expression for ensemble entropy. We shall show later that $S(\hat{\rho})$ coincides with the thermodynamic entropy, assuring us that information entropy and physical entropy are the same. For these reasons, we shall refer to the ensemble entropy, $S(\hat{\rho})$, as the Boltzmann–Gibbs–Shannon (BGS) entropy also.

Entropy as Uncertainty: The Maximum Uncertainty/Entropy Principle

The measurement entropy defined in equation 1 is a good measure of the degree of uncertainty in the measured values of A, and it can also be used to describe the joint uncertainty of two incompatible observables. Indeed, following Deutsch (1983), one can define the joint uncertainty of a pair of measurements to be the sum of their measurement entropies, and proceed to demonstrate that such a measure possesses the correct properties. Furthermore, in many ways the entropic measure of uncertainty proves to be more appropriate than others, particularly in dealing with fundamental questions of measurement theory (Blankenbecler and Partovi 1985; Deutsch 1983; Partovi 1983).

~523~

As an example, consider the problem of describing the state of a quantum system on the basis of (incomplete) information obtained from a series of measurements such as described in the previous section. Clearly, we must demand that the state of the system, described as a density matrix $\hat{\rho}$, should (a) incorporate all the known data, and (b) imply nothing that is not implied by the measured data. Operationally, these conditions are implemented by demanding that $\hat{\rho}$ reproduce the known data and otherwise imply as little else as possible. Thus the answer is obtained by maximizing the BGS entropy subject to the constraints implied by the known data. This quantum principle of maximum uncertainty/entropy, which closely follows Jaynes's maximum entropy principle

(Jaynes 1957), was developed by Blankenbecler and Partovi (1985). A generalization of this formalism to the case where measurements are carried out at different times was developed by Partovi and Blankenbecler (1986) and applied to the longstanding problem of time–energy uncertainty relations. It was then possible to establish unambiguous definitions of the meanings of these relations and achieve rigorous derivations of the corresponding inequalities. These results provide a convincing demonstration of the power and relevance of entropic methods in quantum mechanics.

Quantum Thermodynamics

Are the laws of thermodynamics—equivalently, any of the postulates commonly adopted as the basis of statistical mechanics—independent laws of nature, or do they in fact follow from the underlying dynamics? Ever since Boltzmann's brilliant attempt at deriving thermodynamics from dynamics by means of his H-theorem, there have been countless attempts at resolving this issue (Wehrl 1978). We believe the question has now been settled at the quantum level (Partovi 1989b), and it is our purpose here to define thermodynamics for quantum systems and describe how the zeroth and second laws actually follow from quantum dynamics without any further postulates (the first and third laws are direct consequences of dynamical laws and need not be considered).

As described earlier, an ensemble of similarly prepared quantum systems is described by a density matrix $\hat{\rho}$. Note that an ensemble is not a physical aggregate, and the members of an ensemble must not have any physical influence upon one another. Thus, strictly speaking, a gas of molecules interacting with some other system is not the same thing as an ensemble of molecules interacting with that system; in the latter instance, the molecules are presumed to interact independently, one molecule at a time.

The subject of quantum thermodynamics is interaction between individual quantum systems. Thus as system a, a member of an ensemble described by $\hat{\rho}_a$, interacts with system b, a member of an ensemble described by $\hat{\rho}_b$, one inquires whether the usual thermodynamic laws hold. Just as in macroscopic thermodynamics, one can distinguish a special category of processes which may be characterized as interactions with given, external forces and which may be described by changes in the Hamiltonian operator of the system. Strictly speaking, such descriptions are always approximate, albeit useful ones, and rely on the assumption that the effect of the system on the agent causing the "external force" may be neglected. Thus, in such cases the dynamics of the external agent is ignored and one speaks of exchange of energy in the form of *work*. The general situation, on the other hand, involves the dynamics of both systems and corresponds to *thermal* interactions and the exchange of *heat*.

As usual, we define an ensemble (or a state) to be stationary if it is constant, i.e., if $(\partial/\partial t)\hat{\rho} = 0$. Furthermore, we shall say that a pair of stationary states are in equilibrium if upon interaction they remain unchanged. It is not difficult to show (Partovi 1989b) that in general a pair of states a and b will be in equilibrium if and only if they are of the form $\exp(-\beta_a \hat{H}a)$ and $\exp(-\beta_b \hat{H}_b)$, respectively, with $\beta_a = \beta_b$. Here $\hat{H}$ denotes the Hamiltonian operator for each system (in the absence of the other). These states are known as Gibbs states and play a unique and distinguished role, as will become evident shortly.

Now consider a typical interaction between a member of ensemble a and a member of another, independently prepared ensemble b. In such a situation, what are the chances that the two are correlated before the interaction? Under ordinary circumstances, the answer is essentially zero. Upon interaction, on the other hand, they will in general develop correlations, so that the sum of the *individual* BGS entropies for the two systems will be greater subsequent to the interaction; herein lies the origin of the second law.

To see the connection just asserted, let us first inquire what happens if one of the two systems, say b, is in a Gibbs state before the interaction starts. Using certain general properties of entropy, we find (Partovi 1989b) that the inequality $\Delta S_b - \beta_b \Delta U_b \leq 0$ holds and from this conclude that

$$\Delta S_a - \beta_b \Delta U_a \geq 0. \qquad (3)$$

Here ΔS_a and ΔU_a are the changes in the entropy and energy of system a, and β_b is the parameter characterizing the *initial* Gibbs state of system b. It is important to realize that the inequality in equation 3 is a *non-equilibrium* result, since except for the initial state of system b, all other states (including the final state of b) will in general be non-equilibrium states. Furthermore, there is no implication in equation 3 that the changes in entropy and energy of either system are in any way small. Finally, appearances notwithstanding, the left-hand side of equation 3 is not related to a change in the Helmholtz free energy of system a (a quantity which is only defined for equilibrium states; besides, β_b is a property of the initial state of b and has nothing to do with system a).

The zeroth law can now be obtained from equation 3 by considering both systems a and b to be initially in Gibbs states. Then one has $\Delta S_a - \beta_a \Delta U_a \leq 0$ as well as $\Delta S_b - \beta_b \Delta U_b \leq 0$. These combine to give $\beta_a \Delta U_a + \beta_b \Delta U_b \geq \Delta S_a + \Delta S_b \geq 0$. Since $\Delta U_a + \Delta U_b = 0$ (conservation of energy), one has $(\beta_a - \beta_b)\Delta U_a \geq 0$. This inequality implies that the flow of energy is away from the system with the smaller value of the β parameter. With β identified as inverse temperature, and the property established earlier that Gibbs states with the same value of β do not change upon interaction, we have arrived at the zeroth law of thermodynamics (note that in our units Boltzmann's constant equals unity).

To derive the second law, consider a cyclic change of state for system a brought about by interaction with a number of systems b_i, which are initially in equilibrium at inverse temperatures β_i. Each interaction obeys the inequality in equation 3, so that $\Delta S_{ai} - \beta_i \Delta U_{ai} \geq 0$ for the ith interaction. Since in a cyclic change $\Delta S = \Delta U = 0$, it follows that $\sum_i \Delta S_{ai} = 0$. Summing the inequality stated above on the index i, one arrives at

$$\sum_i \beta_i \Delta U_{ai} \leq 0. \qquad (4)$$

This inequality is a precise statement of the Clausius principle. Note that in conventional terms ΔU_{ai} would be the heat absorbed from system b_i, as explained earlier. Note also that system a need not be in equilibrium at any time during the cycle, and that the β_i only refer to the initial states of the systems b_i.

The Clausius principle established above is equivalent to the second law of thermodynamics, and the entropy function defined from it is none other than the one we have been using, namely the BGS entropy.

Further results on approach to equilibrium, the unique role of the canonical ensemble in quantum thermodynamics, and the calculation of the rate of approach to equilibrium in a specific example can be found in Partovi (1989b).

Reduction and Entropy Increase in Quantum Measurements

In describing the measurement process earlier, we postponed the discussion of what actually happens to those copies of the quantum system that are subjected to interaction with the measuring device. The purpose of this section is to consider the system–device interaction and derive the phenomenon of reduction characteristic of a quantum measurement.

The problem of course is that the evolution of a quantum system during its interaction with a measuring device *appears* to be in violation of the dynamics it is known to obey at other times, a paradox that is known as the *measurement problem.* Indeed, it is customary to postulate that a quantum system does not obey the known dynamics of evolution during a measurement process, thereby disposing of the measurement problem by decree. Many physicists, however, believe that this *reduction* postulate is merely a working model of an underlying process to be uncovered and explained on the basis of known dynamics. Indeed important progress has been made along these lines by Zeh (1971), Zurek (1982), Peres (1986), and others, who have emphasized that the seemingly paradoxical behavior of a quantum system during the act of measurement results from interactions with the environment, i.e., with the countless unobserved degrees of freedom to which the system–device complex is unavoidably coupled. In the following, we shall describe the main elements of a recent analysis (Partovi 1989a) that demonstrates in a general and rigorous manner how interaction with the environment leads to the reduction of the state of the quantum system during the act of measurement.

Recall that a device used to measure an observable A entails a partition of the range of values of A into a number of bins, $\{\alpha_i\}$, and a corresponding decomposition of the Hilbert space generated by the projection operators $\{\hat{\pi}_i^A\}$. Let the state of the quantum system before measurement be described by $\hat{\rho}$. When it is projected onto the eigenmanifolds of the partition, $\hat{\rho}$ appears as $\sum_{i,j} \hat{\pi}_i^A \hat{\rho} \hat{\pi}_j^A$, an expression which may be written as a sum of diagonal and off-diagonal contributions as follows:

$$\hat{\rho} = \sum_i \hat{\pi}_i^A \hat{\rho} \hat{\pi}_i^A + \sum_{i \neq j} \hat{\pi}_i^A \hat{\rho} \hat{\pi}_j^A \equiv \hat{\rho}_R + \hat{\rho}'. \qquad (5)$$

During the measurement process the system interacts with the device, thereby establishing the correlations that will serve to yield the sought-after information by means of a reading of the final state of the measuring device. In symbols, the initial system–device density matrix, $\hat{\Omega} = \hat{\rho}\hat{\Gamma}$ (with $\hat{\Gamma}$ representing the initial state of the device), evolves after a time T into

$$\hat{\Omega}(T) = \sum_i \hat{\pi}_i^A \hat{\rho} \hat{\pi}_i^A \hat{\Gamma}_i + \sum_{i \neq j} \hat{\pi}_i^A \hat{\rho} \hat{\pi}_i^A \hat{\Gamma}_{ij} \equiv \hat{\Omega}_R + \hat{\Omega}'. \qquad (6)$$

Here $\hat{\Gamma}_i$ represents that state of the device which corresponds to the value of $\hat{A}$ turning up in the bin α_i. By contrast, $\hat{\Gamma}_{ij}$ represents a state of the device which corresponds to the state of the system being of the non-diagonal form $\hat{\pi}_i^A \hat{\rho} \hat{\pi}_j^A$. Now in a proper measurement, such non-diagonal contributions are never observed, i.e., $\hat{\Omega}'$ is absent, and all one sees of $\hat{\Omega}(T)$ is the *reduced* part $\hat{\Omega}_R$. This disappearance of the off-diagonal contribution $\hat{\Omega}'$ constitutes the crux of the measurement problem. We will now describe how interaction with the environment in fact serves to eliminate $\hat{\Omega}'$ and leave $\hat{\Omega}_R$ as the final state of the system–device complex.

To establish the result just stated, first we need a theorem on the decay of correlations. Let the *correlation entropy*, C_{AB}, between two systems A and B be defined as the difference $S_A + S_B - S_{AB}$. Note that C_{AB} is non-negative, vanishing only when the two systems are uncorrelated, i.e., when $\hat{\rho}_{AB} = \hat{\rho}_A \hat{\rho}_B$.

Now consider four systems A, B, C, and D, initially in the state $\hat{\rho}_{ABCD}(0) = \hat{\rho}_{AB}(0)\hat{\rho}_C(0)\hat{\rho}_D(0)$. The notation implies that systems A and B are initially correlated while all other pairs are initially uncorrelated. Starting at $t = 0$, system A interacts with system C while system B interacts with system D. Then, using a property of the BGS entropy known as *strong subadditivity* (Lieb and Ruskai 1973), one can show (Partovi 1989a) that $C_{AB}(t) \leq C_{AB}(0)$. In other words, interactions with other systems will in time serve to decrease the correlations initially present between A and B. This intuitively "obvious" result is actually a highly non-trivial theorem that depends on the subadditivity property of entropy, itself a profound property of the BGS entropy.

A measuring device, or more accurately the part of it that directly interacts with the quantum system, has a very large cross-section for interaction with the rest of the universe, or its *environment*. Therefore, although the system–device interaction ceases after the establishment of correlations in $\hat{\Omega}(T)$, the device continues to interact with the environment. According to the result established above, on the other hand, this causes the system–device correlations to decay, so that the final value of the system–device correlation entropy will be the minimum consistent with the prevailing conditions.

A closer examination (Partovi 1989a) of the structure of $\hat{\Omega}(T)$ in equation 6, together with the conditions that the measuring device must obey, reveals that the minimum system–device correlation entropy is reached when $\hat{\Omega}' = 0$, i.e., when $\hat{\Omega}(T)$ is in fact reduced to $\hat{\Omega}_R$, thus establishing the fact that it is the interaction with the environment which brings about the reduction of the state of the system. It is now clear why reduction appears to be totally inexplicable when viewed in the context of system–device interactions only.

The reduction process described above entails an entropy increase, given by $\Delta S = S(\hat{\Omega}_R) - S(\hat{\Omega})$. A straightforward calculation of this entropy increase gives (Partovi 1989a)

$$\Delta S = \text{tr } \hat{\rho} \ln \hat{\rho} - \text{tr} \sum_i \hat{\pi}_i^A \hat{\rho} \ln \hat{\pi}_i^A \hat{\rho}, \qquad (7)$$

with the obvious interpretation that the entropy increase comes about as a result of reducing the initial state of the system $\hat{\rho}$ to the final state $\sum_i \hat{\pi}_i^A \hat{\rho} \hat{\pi}_i^A$, with the off-diagonal elements removed; cf. equation 5.

- 531 -

As an application of equation 7, we will consider the measurement (in one dimension) of the momentum of a system initially in a pure (hence idealized) Gaussian state with a momentum spread equal to μ. The measuring device will be assumed to have uniform bins of size Δp (roughly equal to the resolution of the momentum analyzer). Then one finds from equation 7 that $\Delta S = -\sum_i \mathcal{P}_i \ln \mathcal{P}_i$, where

$$\mathcal{P}_i = (\pi \mu^2)^{-1/2} \int_i dp \, \exp\left(-\frac{p^2}{\mu^2}\right).$$

Here the integral extends over the ith bin. Note that ΔS is precisely what we named *measurement entropy* before.

Consider now the following limiting values of ΔS. For a crude measurement, $\Delta p \gg \mu$, such that practically all events will turn up in one channel (or bin), say, channel k. Then we have $\mathcal{P}_k \cong 1$ and $\mathcal{P}_i \cong 0$ for $i \neq k$, and we find $\Delta S = 0$, exactly as expected. For a high-resolution analyzer, on the other hand, $\Delta p \ll \mu$, so that $\mathcal{P}_i \cong (\pi \mu^2)^{-1/2} \Delta p \exp(-p_i^2/\mu^2)$, and we find

$$\Delta S \cong \frac{1}{2}(3 + \ln \pi) + \ln\left(\frac{\mu}{\Delta p}\right) \qquad (\mu \gg \Delta p). \qquad (8)$$

Thus the entropy increase for reducing the state of the system grows indefinitely as the resolution of the momentum analyzer is increased. Again this is exactly as expected, and points to the impossibility of producing pure states by means of a (necessarily) finite preparation procedure.

It should be pointed out at this point that equation 7 actually represents a lower limit to the amount of entropy increase in a measurement, and that the actual value can be far larger than this theoretical minimum.

Concluding Remarks

In the preceding sections we have described certain basic ideas about the role and meaning of entropy in quantum mechanics, and have outlined a number of applications of these ideas to longstanding problems in quantum theory and statistical mechanics. Among these are the quantum maximum uncertainty/entropy principle, multitime measurements and time–energy uncertainty relations, the reversibility problem of statistical mechanics, and the measurement problem of quantum theory. On the basis of the results obtained so far (the details of which can be found in the original papers cited above), it should be amply clear that entropy, properly defined and applied, is a most powerful notion for dealing with problems of foundation in quantum mechanics. As remarked earlier, this is because the manner in which measurements yield information about a quantum system is unavoidably statistical in nature, thus entailing all the usual consequences of dealing with incomplete information, including entropy.

In retrospect, it is rather remarkable how the dynamics of elementary, microscopic systems of a few degrees of freedom can turn into a statistical problem of considerable complexity when dealing with measured data. ❧

Acknowledgments

This work was supported by the National Science Foundation under grant no. PHY-8513367 and by a grant from California State University, Sacramento.

REFERENCES

Blankenbecler, R., and M. H. Partovi. 1985. "Uncertainty, Entropy, and the Statistical Mechanics of Microscopic Systems." *Physical Review Letters* 54 (5): 373–76. doi:10.1103/physrevlett. 54.373.

Deutsch, D. 1983. "Uncertainty in Quantum Measurements." *Physical Review Letters* 50 (9): 631–33. doi:10.1103/physrevlett.50.631.

Jaynes, E. T. 1957. "Information Theory and Statistical Mechanics." *Physical Review* 106 (4): 620–30. doi:10.1103/physrev.106.620.

Lieb, E. H., and M. B. Ruskai. 1973. "A Fundamental Property of Quantum-Mechanical Entropy." *Physical Review Letters* 30 (10): 434–36. doi:http://dx.doi.org/10.1103/physrevlett.30.434.

Partovi, M. H. 1983. "Entropic Formulation of Uncertainty for Quantum Measurements." *Physical Review Letters* 50 (24): 1883–85. doi:10.1103/physrevlett.50.1883.

———. 1989a. "Irreversibility, Reduction, and Entropy Increase in Quantum Measurements." *Physics Letters A* 137 (9): 445–50. doi:10.1016/0375-9601(89)90222-3.

———. 1989b. "Quantum Thermodynamics." *Physics Letters A* 137 (9): 440–44. doi:10.1016/0375-9601(89)90221-1.

Partovi, M. H., and R. Blankenbecler. 1986. "Time in Quantum Measurements." *Physical Review Letters* 57 (23): 2887–90. doi:10.1103/physrevlett.57.2887.

Peres, A. 1986. "When is a Quantum Measurement?" *American Journal of Physics* 54 (8): 688–92. doi:10.1119/1.14505.

Shannon, C. E. 1948. "A Mathematical Theory of Communication." *Bell System Technical Journal* 27 (4): 623–56. doi:10.1002/j.1538-7305.1948.tb00917.x.

Wehrl, A. 1978. "General Properties of Entropy." *Reviews of Modern Physics* 50 (2): 221–60. doi:10.1103/revmodphys.50.221.

Zeh, H. D. 1971. "On the Irreversibility of Time and Observation in Quantum Theory." In *Foundations of Quantum Mechanics,* edited by B. d'Espagnat. New York, NY: Academic Press.

Zurek, W. H. 1982. "Environment-Induced Superselection Rules." *Physical Review D* 26 (8): 1862–80. doi:10.1103/physrevd.26.1862.

66

EINSTEIN COMPLETION OF QUANTUM MECHANICS MADE FALSIFIABLE

O. E. Rössler, University of Tübingen

- 537 -

An experiment is proposed in which correlated photons are used to obtain information on the relativistic measurement problem. In a long-distance version of the Ou–Mandel experiment, one of the two measuring stations can be put into an orbiting satellite whose position is tightly monitored.

The current view, which holds that in invariant spacetime mutually incompatible interpretations of measured data are possible (so that in effect the commutation relations can be violated), thereby becomes amenable to empirical falsification. Chances are that there will be no surprise. The alternative: a new non-local quantum effect (and a strengthened role of the observer in the sense of Wheeler) may turn out to exist.

The big mystery in the formalism of quantum mechanics is still the measurement problem—the transition from the linear probability amplitude formalism to the nonlinearly projected individual events (Jammer 1974). In contrast, the "relativistic measurement problem," in which everything is compounded by relativistic considerations, has so far even resisted all attempts at formalization.[1]

In the context of the ordinary measurement problem, the paradigm of correlated photons (Aspect, Grangier, and Roger 1982; Aspect, Dalibard, and Roger 1982; Kocher and Commins 1967; Wheeler 1946) has already proven an invaluable

---

[1] See Aharonov and Albert (1981, 1984), Bloch (1967), Landau and Peierls (1931), and Schlieder (1968).

empirical tool. While the two individual projection results remain probabilistic, they nevertheless are strictly correlated across the pair—as if one and the same particle were available twice! Therefore, the question arises of whether the same tool may not be transplanted away from its original domain (that of confirming quantum mechanics) in order to be used as a probe in the unknown terrain of relativistic quantum mechanics.

    A similar proposal has been once made with disappointing results. Einstein (1928) had thought of subjecting two correlated particles to a condition in which both are causally insulated (space-like separated) in order to, at leisure, collect from each the result of a different projected property of the original joint wave function (see also Einstein, Podolsky, and Rosen 1935). Since in this way two incompatible (non-commuting) measurement results could be obtained from the same wave function, his declared aim was to "complete" quantum mechanics in this fashion. To everyone's surprise, Bell (1964) was able to demonstrate that the two particles remain connected "non-locally." They behave exactly as if the distant measurement performed on the first had been performed twice, namely on the second particle, too, in the form of a preparatory measurement. More technically speaking, the distant measurement throws both particles into the same eigenstate (reduction at a distance). The achievement of Bell was to show that this implication of the quantum-mechanical formalism is indeed incompatible with any pre-existing set of properties of the two particles that would make the effect at a distance only an apparent one. A painstaking analysis of all relative angles and their attendant correlations was the key step. Thus, Einstein's intuition was proven wrong for once. No more than one "virgin reduction" of the original wave function need be assumed.

However, the mistake made by Einstein may have been smaller than meets the eye. His specific proposal to use relativistic insulation (space-like separation) as a means to "fool" quantum mechanics was presumably chosen for didactic reasons only. The larger idea—to use relativity theory for the same purpose—is still unconsummated. There may exist a second mechanism at causal separation between two space-like separated events that when applied to the two measurements might indeed "decouple" them, so that quantum mechanics could be fooled indeed—or else would have to respond with an even more vigorous and surprising defense.

~539~

Such a second mechanism, in fact, exists, as is well known. The temporal ordering between two space-like separated events (their causal relationship, so to speak) is not a relativistic invariant. The very "connection" discovered by Bell makes this result, which ordinarily poses no threat to causality (Zeeman 1964), conducive to carrying an unexpected power.

Let us illustrate the idea in concrete terms (fig. 1). The two measuring stations used in the Aspect experiment here are assumed to have been put in motion relative to each other (Aspect, Grangier, and Roger 1982; Aspect, Dalibard, and Roger 1982). Moreover, the two distances are chosen so carefully that exactly the above condition (reversal of priority relations as to which measuring device is closer to the point of emission in its own frame) is fulfilled. In consequence, each *half* experiment is identical to an ordinary Aspect experiment in which the most important measurement (the first) has already taken place. Only after this first reduction has been obtained in the frame in question will there be a second measurement. This second measurement, of course, will be performed by a moving (receding) measurement device. But since by that time the joint reduction of the photon's state has already been accomplished,

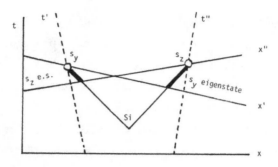

Figure 1. Lorentz diagram of the proposed relativistic experiment using correlated photons. The two measuring devices (dashed lines) are situated in two different frames (primed and double-primed) that are receding from each other. The singlet-state-like wave function of the photon pair (Si) is subject to two primary reductions valid simultaneously in the third (unprimed) frame. See text; e.s. = eigenstate.

the other photon already possesses a well-defined spin. Hence, by the well-known fact that a photon's spin yields the same fixed outcome whatever the state of the head-on motion of the measuring device (Bjorken and Drell 1964), there is indeed no difference relative to an ordinary Aspect experiment. A "catch-22" situation has therefore been achieved.

The question of how nature will respond to this experiment is open. Let us, therefore, first check whether the experiment can, in fact, be done (fig. 2). Here for simplicity only two frames are assumed, one stationary and the other moving. Both are completely symmetric. If, as shown, detector delays of 3 nanoseconds ($d = 1$ light meter) are assumed, and if, in addition, satellite velocity ($w = 11\,\mathrm{km/sec}$) is assumed for the second frame so that $v/c = 4 \times 10^{-5} \ll 1$, one sees from the diagram that $s = d \times c/v = 2.5 \times 10^4\,\mathrm{m} = 25\,\mathrm{km} = 16\,\mathrm{mi}$. This amounts to a rather "long-distance" version of the Aspect experiment. The weak intensity of the source used in the latter (Aspect, Grangier, and Roger 1982; Aspect, Dalibard, and Roger 1982) would certainly forbid such an extension. Moreover, the two photons are not simultaneously emitted in this experiment

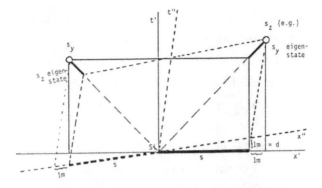

Figure 2. The experiment of figure 1, redrawn in more detail in the two frames x' and x''. Note that the slope of the x'' axis equals v/c, but also is equal to d/s (with d measured in meters) as long as v is much smaller than c. With d and v fixed, s (the minimum distance to the source from either measuring device in its own frame) can, therefore, be calculated. Compare text.

(Aspect, Grangier, and Roger 1982; Aspect, Dalibard, and Roger 1982; Kocher and Commins 1967). Therefore, it is fortunate that a more recent experiment exists which is *both* high-intensity *and* of the simultaneously emitting type since the two photons are generated in parametric down conversion before being superposed (Ou and Mandel 1988). Therefore, the present experiment can be actually implemented in two steps, by first scaling up the Ou–Mandel experiment and then making one of the two measuring devices (analyzer plus detector) spacebound.

This essentially concludes the message of the present note. What remains is to make the connection to other work. While the present experiment is new, Shimony (1986) recently looked at a rather similar case. His mathematical analysis (done without pictures) fits in perfectly as a complement to the present context. The only difference: he did not differentiate between the two measuring devices being mutually at rest or not. He, therefore, could rely entirely on the Bell experiment, with his added analysis only having the character of a *gedankenexperiment* that cannot and need not be done since all the facts are available anyhow. His conclusion nevertheless was quite revolutionary since it culminated in the conjecture that the quantum-mechanical notion of a measured eigenstate may have to be redefined such that it becomes frame-dependent.

Shimony's conclusion had been reached before by Schlieder (1968) and Aharonov and Albert (1984). These authors applied a well-known axiom from relativistic quantum mechanics (that two space-like separated measurements always commute; see Schlieder [1968]) to correlated particles, arriving at the theorem that the same particle may possess multiple quantum states (density matrices) at the same point in spacetime. Specifically, these states form—in accordance with earlier proposals of Dirac, Tomonaga, and Schwinger—a "functional" on the set of space-like hypersurfaces intersecting the point, such that on each hypersurface the rules of non-relativistic quantum mechanics are obeyed (Aharonov and Albert 1984). Thus, a new "selection rule" is invoked which picks the admissible interpretation out of a set of invariant empirical facts. Since these facts imply what can be called an "overdetermination" in spacetime relative to what naive quantum mechanics permits, Einstein's prediction appears to be vindicated. An alternative interpretation, given by Park and Margenau (1971), is even more fitting. These authors proposed to acknowledge the weakened status of the commutation relations by modifying one of the axioms of quantum mechanics used in von Neumann's simultaneous measurability theorem ("weak" rather than strong correspondence between linear Hermitian operators on Hilbert space and complete orthonormal sets of eigenvectors, on the one hand, and physical observables, on the other; see von Neumann [1955]).

However, after the unsuspected success of Bell in keeping up the status of the commutation relations in the *non-*

relativistic realm, it is a legitimate question to ask whether or not one can be absolutely sure that Einstein's "improved" (relativistic) proposal is in accordance with reality. Specifically, can the paradoxical fact that from the point of view of relativistic quantum mechanics the Aspect experiment implies exactly the opposite of what it does in non-relativistic quantum mechanics—namely, that the commutation relations are *not* respected in spacetime—perhaps be subjected to empirical scrutiny?

-543-

Now the experiment of figure 1 becomes of interest. It is "stronger" than the ordinary Aspect experiment. That the latter opinion may already suffice to obtain a violation of the commutation relations in the relativistic realm (Shimony 1986) is subject to the objection that in the rest frame of the two measuring devices no violation occurs and that, therefore, no violation can be inferred for by-flying frames which only pick up the macroscopic events (induced light flashes, say) that occur in that frame—for these particular macroscopic events cannot have any deeper consequences concerning causality than any other pair of macroscopic events has. Moreover, even if the assertion of an exception existing were correct, it would not be verifiable since nothing new can be learned by definition from the whole set-up as only known data enter it. These objections do not apply to the experiment of figure 1. It (a) is critically different from all data recorded previously and (b) certainly implies a violation of the commutation relations if its outcome is unchanged compared to the ordinary Aspect experiment, since, as shown above, two mutually incompatible Aspect experiments are combined in it according to Bell's non-relativistic theory.

Chances are, of course, that the current opinion that relativistic invariance (unlike relativistic locality) has precedence over the commutation relations can be upheld after the experiment has been performed, because its outcome will be negative: no change in correlations relative to the standard Aspect experiment. However, now there is a tiny bit of a chance that, should something be profoundly wrong with our opinion about nature, this fact will make itself felt *somehow* as one "save" the commutation relations (by excluding joint reductions) and spell the end of the doctrine of an observer-invariant spacetime (since the state of motion of a measuring device could affect photon spin). However, it would be much too "heavy" to be seriously proposed as a prediction. All quantitative theory available would thereby be contradicted. There is nothing on the horizon that could seriously threaten the *invariance of quantum spacetime*. What is new is only that the latter has become empirically confirmable (and therefore also "in principle falsifiable") for the first time.

To conclude, a new quantum experiment feasible with current technology has been proposed. The status of the commutation relations in the relativistic measurement problem can be decided. Specifically, the "space-borne" Ou–Mandel experiment will show whether (1) the current idea that there exists an observer-invariant quantum spacetime can be upheld (Einstein completion) or (2) the observer is reinforced in his "participatory" (Wheeler 1977) role in a whole new context. ⚡

Acknowledgments

I thank Wojciech Zurek and Jens Meier for discussions and John Bell for a correction concerning figure 2.

Added in proof: Peres (1984) used a diagram similar to figure 1, which was redrawn for this paper.

REFERENCES

Aharonov, Y., and D. Z. Albert. 1981. "Can We Make Sense Out of the Measurement Process in Relativistic Quantum Mechanics?" *Physical Review D* 24 (2): 359–70. doi:10.1103/physrevd.24.359.

———. 1984. "Is the Usual Notion of Time Evolution Adequate for Quantum-Mechanical Systems? II. Relativistic Considerations." *Physical Review D* 29 (2): 228–34. doi:10.1103/physrevd.29.228.

Aspect, A., J. Dalibard, and G. Roger. 1982. "Experimental Test of Bell's Inequalities Using Time-Varying Analyzers." *Physical Review Letters* 49 (25): 1804–07. doi:10.1103/physrevlett.49.1804.

Aspect, A., P. Grangier, and G. Roger. 1982. "Experimental Realization of Einstein-Podolsky-Rosen-Bohm *Gedankenexperiment*: A New Violation of Bell's Inequalities." *Physical Review Letters* 49 (2): 91–94. doi:10.1103/physrevlett.49.91.

Bell, J. S. 1964. "On the Einstein Podolsky Rosen Paradox." *Physics Physique Fizika* 1 (3): 195–200. doi:10.1103/physicsphysiquefizika.1.195.

Bjorken, J. D., and S. D. Drell. 1964. *Relativistic Quantum Mechanics*. New York, NY: McGraw-Hill.

Bloch, I. 1967. "Some Relativistic Oddities in the Quantum Theory of Observation." *Physical Review* 156 (5): 1377–84. doi:10.1103/physrev.156.1377.

Einstein, A. 1928. In *Institut International de Physique Solvay, Rapport et Discussions du 5ᵉ Conseil,* 253. Paris, France.

Einstein, A., B. Podolsky, and N. Rosen. 1935. "Can Quantum-Mechanical Description of Physical Reality Be Considered Complete?" *Physical Review* 47 (10): 777–80. doi:10.1103/physrev.47.777.

Jammer, M. 1974. *The Philosophy of Quantum Mechanics: The Interpretations of Quantum Mechanics in Historical Perspective*. New York, NY: Wiley.

Kocher, C. A., and E. D. Commins. 1967. "Polarization Correlation of Photons Emitted in an Atomic Cascade." *Physical Review Letters* 18 (15): 575–77. doi:10.1103/physrevlett.18.575.

Landau, L., and R. Peierls. 1931. "Erweiterung des Unbestimmtheitsprinzips für die relativistische Quantentheorie." *Zeitschrift für Physik* 69 (1–2): 56–69. doi:10.1007/bf01391513.

Ou, Z. Y., and L. Mandel. 1988. "Violation of Bell's Inequality and Classical Probability in a Two-Photon Correlation Experiment." *Physical Review Letters* 61 (1): 50–53. doi:10.1103/physrevlett.61.50.

Park, D., and H. Margenau. 1971. "The Logic of Noncommutability of Quantum-Mechanical Operators—and Its Empirical Consequences." In *Perspectives in Quantum Theory*, edited by W. Yourgrau and A. Van der Merwe, 37–70. Boston, MA: MIT Press.

Peres, A. 1984. "What Is a State Vector?" *American Journal of Physics* 52 (7): 644–50. doi:10.1119/1.13586.

Schlieder, S. 1968. "Einige Bemerkungen zur Zustandsänderung von relativistischen quantenmechanischen Systemen durch Messungen und zur Lokalitätsforderung." *Communications in Mathematical Physics* 7 (4): 305–31. doi:10.1007/bf01646663.

Shimony, A. 1986. "Events and Processes in the Quantum World." In *Quantum Concepts in Space and Time,* edited by R. Penrose and C. J. Isham, 182, 193–95. Oxford, UK: Clarendon Press.

von Neumann, J. 1955. "Simultaneous Measurability and Measurability in General." In *Mathematical Foundations of Quantum Mechanics,* 225–30. Princeton, NJ: Princeton University Press.

Wheeler, J. A. 1946. "Polyelectrons." *Annals of the New York Academy of Sciences* 48 (3): 219–38. doi:10.1111/j.1749-6632.1946.tb31764.x.

———. 1977. "Genesis and Observership." In *Foundational Problems in Special Sciences,* edited by R. E. Butts and K. J. Hintikka. Dordrecht, Netherlands: Reidel.

Zeeman, E. C. 1964. "Causality Implies the Lorentz Group." *Journal of Mathematical Physics* 5 (4): 490–93. doi:10.1063/1.1704140.

$6\dashv$

QUANTUM MECHANICS AND ALGORITHMIC COMPLEXITY

J. W. Barrett, The University, Newcastle upon Tyne

~549~

Although "algorithmic complexity" is in the title of this contribution, I think the subject of this paper is really why it is so hard to formulate the notion of algorithmic information for a quantum system, in an intrinsically quantum manner. Computation is a classical process, and bits are classical facts.

I am hoping to toss out some questions, and hopefully my remarks will provoke some thought. However, I don't claim to be providing the answers. When I gave the talk, I said that there was one "complaint" on every transparency: these are complaints I have about our current theory, and frustration with the fact that we understand the quantum so incompletely.

Quantum Mechanics

One fairly standard view of quantum mechanics is the following (see the contributions by Omnès and Gell-Mann in this volume): after irreversible coupling to the environment, the properties of an object are consistent with the idea that one of the alternatives for its behavior has occurred in a definite way. The clause "in a definite way" is important here. I also want to stress that I am saying that the properties of an object become consistent with the idea that something definite has occurred, but no further; I am not pinning down a time at which one alternative is chosen.

I would like to consider a universe consisting of a small number of finite quantum systems, which we think of as exchanging information. There is a difficulty, because if one system gains information about a second one through an interaction, there is the possibility that the interaction might be "undone" later on, by some interaction which makes use of the quantum correlations between the two systems. Thus, the information gained has only a definite character so long as we forgo the further "use" of some of the possible quantum interactions. This definiteness which I am talking about is, I think, captured by the idea of a consistent logic, introduced by Omnès and having its roots in Griffiths's idea of a consistent history for a quantum system (Griffiths 1987).

In the real world, there is always a large-scale, decohering environment involved when we, human beings, gain information: our own bodies, if nothing else. In most cases, though, it is an inanimate part of the environment. Then, it becomes effectively impossible to show up the coherence between the small system and the macroscopic object, because the experimental manipulations involved get too complicated. They would involve a huge amount of information. Incidentally, this shows up some sort of link between information, of an algorithmic kind (how to program a robot doing the experiments), and the measure of correlation information discussed by Everett which uses the density matrix (DeWitt and Graham 1973).

Because the environment "decoheres" interactions in the real world, the present discussion is optional from a strictly practical viewpoint. However, I am unhappy with this resolution of the problem and think that quantum mechanics ought to make sense without the necessary inclusion of such very large systems.

Bell's Analysis and Set-Theoretic Assumptions

In the two-spin experiment of Einstein, Podolsky, Rosen, and Bell fame, one can understand Bell's result by saying that one has four sets of probabilities for the outcomes of each of the four possible mutually incompatible experiments (the four experiments relating to the possibility of setting the analyzer to each of two different angles at the two different locations), but that there is no overall joint probability distribution for all sixteen possible outcomes $O = \{(u,u,u,u), (u,u,u,d), \ldots, (d,d,d,d)\}$. The quadruple (u,u,u,d), for example, refers to the hypothesis that the left-hand spin would be up with either orientation of its polarizer, while the right-hand spin would be up with one orientation and down with the other one. This point of view is, I think, due to Wigner. In arriving at this conclusion, one uses the usual rules of probability theory; namely, one forms the marginal distribution in the two variables measured, by summing over the probabilities with different values of the unobserved variables.

I would like to stress the following interpretation: there is really no underlying set theory for the probabilities which one has; the set O does not exist. One cannot say that there is an exhaustive collection of properties of the system which define it as one element of a particular set of all the possibilities.

Whilst I am discussing the Bell analysis, I would like to mention the possibility of a computer science version of the result. This came up in the questions to my talk.

Suppose the universe is a cellular automaton of finite size and that it contains two-spin experiments, analyzers and all. Suppose, also, that the way the machine works is that each analyzer is set to one of two possible angles according to the value of a bit (0 or 1) in one particular square close (in space and time) to the analyzer (the "setting bit"). This bit is in such a position that, due to the finite speed of propagation of computational "light," its value cannot influence the outcome of the detection experiment of the other spin. Now run the automaton, but instead of allowing the setting bits to be calculated by the usual rule table from its nearest neighbors, insert, by hand as it were, one bit from a truly random sequence of digits into the setting bit position each time the detection experiment is about to be performed. Whilst such a thing is not a computable automaton, because of the random, rather than pseudorandom, sequence, it is still a perfectly well-defined mathematical object. I think that one would find that, in the long run, if one compiled the frequencies into probabilities, they would not be able to break Bell's inequality.

The reason for this is that although in the short term the left-hand analyzer might be able to successfully "guess," i.e., compute, the right-hand analyzer setting and so provide outcomes to the experiment which are, for example, compatible with quantum mechanics, in the long run this strategy would definitely fail, because one cannot compute more than some definite initial number of digits of an algorithmically random sequence. Thus, the automaton would have to fall back on purely probablistic methods of providing experimental outcomes, being prepared for either analyzer setting, and the results would be in line with the above probablistic reasoning, giving satisfaction of Bell's inequality.

I don't have a formal proof of the above line of reasoning, so strictly speaking it remains a conjecture. There are some assumptions being made; for example, as Ed Fredkin pointed out in his talk, I am assuming the records of the results obtained can be freely propagated to a common location in the automaton and combined in any desired logical combination.

The role of set theory, if I may be allowed to interpret a mathematical formalism in a physical setting, is to provide a systematic way of defining objects by their properties. As John Wheeler might like it, set theory answers the question: what does it mean, "to be"? For example, one of the axioms ("comprehension") asserts that if one has a set of objects and a proposition, then this defines a second set: the subset on which the proposition is true. If one reinterprets this in terms of physical measurements on a system, the set of objects is the set of possible configurations, and the truth value of a proposition is the outcome of a measurement interaction. Reasoning in this very literal way, one is bound to regain the idea of the set O for the two-spin experiment and not be able to reconstruct quantum mechanics. I think other axioms of set theory have an interpretation in terms of physical operations based on a pre-quantum understanding of physics.

Thus we have been accustomed to abandoning the goal of understanding quantum objects entirely in terms of classical set-theoretic constructions, but speak about them in roundabout ways. This is the source of the tension in debates about quantum theory. Omnès has clarified exactly to what extent one can use set-theoretic constructs in quantum theory in a direct way, and where the inconsistencies set in. To my mind this is very important advance. However, I feel that there ought to be a set-theoretic language which applies directly to all quantum interactions. Perhaps it is along the lines Finkelstein (1989) has suggested.

The Quantum Surveyor

Let us move to the question of pinning down the actual information in a quantum two-state spin system. How many bits of information are there in a single system (for example, a single photon "traveling between" polarizer and analyzer)? The idea of a bit is itself straight out of classical set theory, the definite and unambiguous assignment of an element of the set $\{0, 1\}$, and so the assignment of an information content to the photon itself is fraught with difficulties. However, one has a feeling (see the contribution by Schumacher in this volume) that each photon spin cannot convey more than one bit of information.

The quantum surveyor is a surveyor who finds herself caught out in the field with nothing to measure the angle between two distant trees on the horizon but a polarizer, analyzer, photon source, and detector. She points the axes of the polarizer and analyzer in the directions of the two trees and counts the number of individual photons which are transmitted through the polarizer and analyzer, as a fraction of the number which pass the polarizer but are stopped by the analyzer. According to quantum theory, the inverse cosine of the square root of this fraction converges, as the number of photons tends to infinity, to the angle. Clearly one needs to use a large number of photons if one wants to resolve small angles. W. Wootters discussed in this conference the question of how much information one gains about the angle with a finite number of photons.

Clearly, from the point of view of information, there are three separate things here: the information contained in the classical angle, the information in the wavefunction, and the single quantum "bit" of information for an individual photon's spin. Remember that from the point of view of this talk, I don't want to take the macroscopically large for granted; therefore I would consider this experiment the determination of the value of the angle, rather than obtaining some partial information about some pre-existing value from the continuum of real numbers. So, a question arises: are all spacetime measurements ultimately of this type (if usually disguised because of the law of large numbers)? This is the type of question which Penrose (1971) raised when he invented spin networks.

Keeping the macroscopic objects a finite size has other effects. The angle is effectively a property of the spatial relationship between the polarizer and analyzer. A finite size for these means a cutoff in the spectrum of the angular momentum for each object, and hence some uncertainty in the relative angle between the two due to the quantum uncertainty principle. Thus the bit string that one gets in this case, from writing 0 when a photon fails to pass through and 1 if it does pass through, does not define an angle in the classical sense. What I mean is that there is not a continuum range of values which the angle, as defined by the quantum "surveying," can be said with certainty to take.

Thus we see that the continuum nature of spacetime, the continuum nature of the space of quantum wavefunctions, and the usual assumption of the existence of infinitely large and massive reference bodies are inextricably linked. In particular, we see that the quantum wavefunction is not just a property of the photon spin. It is a property of the spacetime measurement as well as of the photon itself.

The implications of this for understanding the concept of information, in an algorithmic sense, in quantum theory are something one cannot ignore. If one wants to deal with a finite amount of information, one has to use systems of a finite size throughout; then one cannot use continuum concepts such as a wavefunction in the conventional sense. I feel that a satisfying resolution of this problem should also be one that solves the puzzles I outlined earlier about the relationship of quantum properties to classical sets. ✌

REFERENCES

DeWitt, B. S., and N. Graham. 1973. *The Many-Worlds Interpretation of Quantum Mechanics.* Princeton, NJ: Princeton University Press.

Finkelstein, D. 1989. "Quantum Net Dynamics." *International Journal of Theoretical Physics* 28 (4): 441–67. doi:10.1007/bf00673296.

Griffiths, R. B. 1987. "Correlations in Separated Quantum Systems: A Consistent History Analysis of the EPR Problem." *American Journal of Physics* 55 (1): 11–17. doi:10.1119/1.14965.

Penrose, R. 1971. "Angular Momentum: An Approach to Combinatorial Space-Time." In *Quantum Theory and Beyond,* edited by T. Bastin. London, UK: Cambridge University Press.

PROBABILITY IN QUANTUM THEORY

E. T. Jaynes, Washington University, St. Louis

For some sixty years, it has appeared to many physicists that probability plays a fundamentally different role in quantum theory than it does in statistical mechanics and analysis of measurement errors. A common notion is that probabilities calculated within a pure state have a different character than the probabilities with which different pure states appear in a mixture or density matrix. As Pauli put it, the former represents ". . . eine prinzipielle Unbestimmtheit, nicht nur Unbekanntheit." But this viewpoint leads to so many paradoxes and mysteries that we explore the consequences of the unified view—all probability signifies only human information. We examine in detail only one of the issues this raises: the reality of zero-point energy.

Introduction: How We Look at Things

In this workshop we are venturing into a smoky area of science where nobody knows what the real truth is. Such fields are always dominated by the compensation phenomenon: supreme self-confidence takes the place of rational arguments. Therefore, we shall try to avoid dogmatic assertions, and only point out some of the ways in which quantum theory would appear different if we were to adopt a different viewpoint about the meaning of probability. We think that the original viewpoint of James Bernoulli and Laplace offers some advantages today in both conceptual clarity and technical results for currently mysterious problems.

How we look at a theory affects our judgment as to whether it is mysterious or irrational on the one hand, or satisfactory and reasonable on the other. Thus, it affects the direction of our research efforts, and *a fortiori* their results. Indeed, whether we theorists can ever again manage to get ahead of experiment will depend on how we choose to look at things, because that determines the possible forms of the future theories that will grow out of our present ones. One viewpoint may suggest natural extensions of a theory, which cannot even be stated in terms of another. What seems a paradox from one viewpoint may become a platitude from another.

For example, a hundred years ago, a much discussed problem was how material objects can move through the ether without resistance. Yet, from another perspective, the mystery disappeared without any need to dispense with the ether. One can regard material objects not as impediments to the "flow" of ether, but as *parts* of the ether ("knots" in its structure) which are propagating through it. With this way of looking at it, there is no mystery to be explained. As a student at Princeton many years ago, I was fascinated to learn from John Wheeler how much of physics can be regarded as simply geometry, in this way.

Today we are beginning to realize how much of all physical science is really only *information*, organized in a particular way. But we are far from unraveling the knotty question: "To what extent does this information reside in us, and to what extent is it a property of Nature?" Surely, almost every conceivable opinion on this will be expressed at this workshop.

Is this variability of viewpoint something to be deplored? Eventually we should hope to present a unified picture to the rest of the world. But for the moment this is neither possible nor desirable. We are all looking at the same reality and trying to understand what it is. But we could never understand the structure of a mountain if we looked at it only from one side. The reality that we are studying is far more subtle than a mountain, and so it is not only desirable but *necessary* that it be examined from many different viewpoints, if we are ever to resolve the mystery of what it is. Here we present one of those viewpoints.

First, we note a more immediate example of the effect of perspective, in order to support, by physical arguments, later suggestions from probability considerations.

How Do We Look at Gravitation and Quantum Electrodynamics?

In teaching relativity theory, one may encounter a bright student who raises this objection: "Why should such a fundamental thing as the metric of space and time be determined only by gravitational fields—the weakest of all interactions? This seems irrational." We explain a different way of looking at it, which makes the irrationality disappear: "One should not think of the gravitational field as a kind of pre-existing force which 'causes' the metric; rather, the gravitational field is the main observable *consequence* of the metric. The strong interactions have not been ignored, because the field equations show that the metric is determined by all the energy present." According to the first viewpoint, one might think it a pressing research problem to clear up the mystery of why the metric depends only on gravitational forces. From the second viewpoint, this problem does not exist.

If the student is very bright, he will be back the next day with another criticism: "If the gravitational field is only a kind of bootstrap effect of the other forces, it raises the question of whether the gravitational field should be quantized separately. Wouldn't we be doing the same thing twice?" Thus different ways of looking at what a gravitational field is might lead one to pursue quite different lines of research.

A similar issue arises in electrodynamics, making a thoughtful person wonder why we quantize the electromagnetic (EM) field. The following observations were made by Albert Einstein in two lectures at the Institute for Advanced Study, which I was privileged to attend in the late 1940s.

He noted that, in contemporary quantum theory, we first develop the theory of electrons via the Schrödinger equation, and work out its consequences for atomic spectra and chemical bonding, with great success. Then we develop the theory of the free quantized EM field independently, and discuss it as a separate thing. Only at the end do we, almost as an afterthought, decide to couple them together by introducing a phenomenological coupling constant 'e' and call the result "quantum electrodynamics."

Einstein told us: "I feel that it is a delusion to think of the electrons and the fields as two physically different, independent entities. Since neither can exist without the other, there is only one reality to be described, which happens to have two different aspects; and the theory ought to recognize this from the start instead of doing things twice."

Indeed, the solution of the EM field equations is, in either classical or quantum theory,

$$A_\mu(x) = \int D(x - y) J_\mu(y) \, d^4 y. \qquad (1)$$

In quantum theory $A_\mu(x)$ and $J_\mu(y)$ are operators, but, since the propagator $D(x - y)$ is a scalar function, the $A_\mu(x)$ in equation (1) is not an operator on a "Maxwell Hilbert space" of a quantized EM field—it is an operator on the same space as $J_\mu(y)$, the "Dirac Hilbert space" of the electrons.

Conventionally, one says that equation (1) represents only the "source field" and we should add to this the quantized "free field" $A_\mu^{(0)}(x)$ which operates on the Maxwell Hilbert space. But, fundamentally, every EM field is a source field from somewhere; therefore, it is *already* an operator on the space of perhaps distant sources. So why do we quantize it again, thereby introducing an infinite number of new degrees of freedom for each of an infinite number of field modes?

One can hardly imagine a better way to generate infinities in physical predictions than by having a mathematical formalism with $(\infty)^2$ more degrees of freedom than are actually used by Nature. The issue is: Should we quantize the matter and fields separately, and then couple them together afterward, or should we write down the full classical theory with both matter and field and with the field equations in integrated form, and quantize it in a single step? The latter procedure (assuming that we could carry it out consistently) would lead to a smaller Hilbert space.

~563~

The viewpoint we are suggesting is quite similar in spirit to the Wheeler–Feynman electrodynamics, in which the EM field is not considered to be a "real" physical entity in itself, but only a kind of information storage device. That is, the present EM field is a "sufficient statistic" that summarizes all the *information* about past motion of charges that is relevant for predicting their future motion.

It is not enough to reply that "the present quantum electrodynamics procedure must be right because it leads to several very accurate predictions: the Lamb shift, the anomalous moment, etc." To sustain that argument, one would have to show that the quantized free field actually plays an essential role in determining those accurate numbers (e.g., 1058 MHz). But their calculation appears to involve only the Feynman propagators; mathematically, the propagator $D(x - y)$ in equation (1) is equally well a Green's function for the quantized or unquantized field.

The conjecture suggests itself, almost irresistibly, that those accurate experimental confirmations of quantum electrodynamics come from the local source fields, which are coherent with the local state of matter. This has been confirmed in part by the "source-field theory" that arose in quantum optics about fifteen years ago (Allen and Eberly 1975; Milonni, Ackerhalt, and Smith 1973; Senitzky 1973). It was found that, at least to lowest non-vanishing order, observable effects such as spontaneous emission and the Lamb shift can be regarded as arising from the source field which we had studied already in classical EM theory, where we called it the "radiation reaction field." Some equations illustrating this in a simpler context are given below.

In these quantum optics calculations, the quantized free field only tags along, putting an infinite uncertainty into the initial conditions (that is, a finite uncertainty into each of an infinite number of field modes) and thus giving us an infinite "zero-point energy" but not producing any observable electrodynamic effects. One wonders, then: *Do we really need it?*

How Do We Look at Basic Quantum Theory?

Current thinking about the role of information in science applies to all areas, and in particular to biology, where perhaps the most valuable results will be found. But the most tangled area in present physical science is surely the standard old 1927 vintage quantum theory, where the conceptual problems of the "Copenhagen interpretation" refuse to go away, but are brought up for renewed discussion by every new generation (much to the puzzlement, we suspect, of the older generation who thought these problems were all solved). Starting with the debates between Bohr and Einstein over sixty years ago, different ways of looking at quantum theory persist in making some see deep mysteries and contradictions in need of resolution, while others insist that there is no difficulty.

Defenders of the Copenhagen interpretation have displayed a supreme self-confidence in the correctness of their position, but this has not enabled them to give the rest of us any rational explanations of why there is no difficulty. Richard Feynman at least had the honesty to admit, "Nobody knows how it can be that way."

We doubters have not shown so much self-confidence; nevertheless, all these years it has seemed obvious to me—for the same reasons that it did to Einstein and Schrödinger— that the Copenhagen interpretation is a mass of contradictions and irrationality and that, while theoretical physics can of course continue to make progress in the mathematical details and computational techniques, there is no hope of any further progress in our basic understanding of Nature until this conceptual mess is cleared up.

Let me stress our motivation: if quantum theory were not successful pragmatically, we would have no interest in its interpretation. It is precisely *because* of the enormous success of the quantum mechanics (QM) mathematical formalism that it becomes crucially important to learn what that mathematics means. To find a rational physical interpretation of the QM formalism ought to be considered the top priority research problem of theoretical physics; until this is accomplished, all other theoretical results can only be provisional and temporary.

This conviction has affected the whole course of my career. I had intended originally to specialize in quantum electrodynamics, but this proved to be impossible. Whenever I look at any quantum-mechanical calculation, the basic craziness of what we are doing rises in my gorge and I have to try to find some different way of looking at the problem that makes physical sense. Gradually, I came to see that the foundations of probability theory and the role of human information have to be brought in, and so

I have spent many years trying to understand them in the greatest generality.

The failure of quantum theorists to distinguish in calculations between several quite different meanings of "probability," between expectation values and actual values, makes us do things that are unnecessary and fail to do things that are necessary. We fail to distinguish in our verbiage between prediction and measurement. For example, two famous vague phrases—"It is impossible to *specify* . . ." and "It is impossible to *define* . . ."—can be interpreted equally well as statements about prediction or about measurement. Thus, the demonstrably correct statement that the present theory cannot *predict* something becomes twisted into the almost certainly false claim that the experimentalist cannot *measure* it!

We routinely commit the Mind Projection Fallacy of projecting our own thoughts out onto Nature, supposing that creations of our own imagination are real properties of Nature, or that our own ignorance signifies some indecision on the part of Nature. This muddying up of the distinction between reality and our knowledge of reality is carried to the point where we find some asserting the objective reality of probabilities while denying the objective reality of atoms! These sloppy habits of language have tricked us into mystical, pre-scientific standards of logic, and leave the meaning of any QM result simply undefined. Yet we have managed to learn how to calculate with enough art and tact so that we come out with the right numbers!

The main suggestion we wish to make is that how we look at basic probability theory has deep implications for the Bohr–Einstein positions. Only within the past year has it appeared to the writer that we might be able finally to resolve these matters in the happiest way imaginable: a reconciliation of the views of

Bohr and Einstein in which we can see that they were both right in the essentials, but just thinking on different levels.

Einstein's thinking is always on the ontological level traditional in physics, trying to describe the realities of Nature. Bohr's thinking is always on the epistemological level, describing not reality but only our information about reality. The peculiar flavor of his language arises from the absence of words with any ontological import; the notion of a "real physical situation" was just not present and he gave evasive answers to questions of the form, "What is really happening?" Eugene Wigner (1974) was acutely aware of and disturbed by this evasiveness when he remarked:

> These Copenhagen people are so clever in their use of language that, even after they have answered your question, you still don't know whether the answer was "yes" or "no"!

J. R. Oppenheimer, more friendly to the Copenhagen viewpoint, tried to explain it in his lectures in Berkeley in the 1946–47 school year. Oppy anticipated multiple-valued logic when he told us:

> Consider an electron in the ground state of the hydrogen atom. If you ask, "Is it moving?" the answer is "no." If you ask, "Is it standing still?" the answer is "no."

Those who, like Einstein (and, up till recently, the present writer), tried to read ontological meaning into Bohr's statements were quite unable to understand his message. This applies not only to his critics but equally to his disciples, who undoubtedly embarrassed Bohr considerably by offering such exegeses as, "Instantaneous quantum jumps are real physical events," or "The variable is created by the act of measurement," or the remark of Pauli quoted above, which might be rendered loosely as, "Not only are you and I ignorant of x and p; Nature herself does not know what they are."

Critics who tried to summarize Bohr's position sarcastically as, "If I can't measure it, then it doesn't exist!" were perhaps closer in some ways to his actual thinking than were his disciples. Of course, while Bohr studiously avoided all assertions of "reality," he did not carry this to the point of denying reality; he was merely silent on the issue, and would prefer to say, simply, "If we can't measure it, then we can't use it for prediction."

Although Bohr's whole way of thinking was very different from Einstein's, it does not follow that either was wrong. In the writer's view, all of Einstein's thinking—in particular the Einstein–Podolsky–Rosen (EPR) argument—remains valid today, when we take into account its ontological character. But today, when we are beginning to consider the role of information for science in general, it may be useful to note that we are finally taking a step in the epistemological direction that Bohr was trying to point out sixty years ago.

This statement applies only to the general philosophical position that the role of human information in science needs to be recognized and taken into account explicitly. Of course, it does not mean that every technical detail of Bohr's work is to remain unchanged for all time. Our present QM formalism is a peculiar mixture describing in part laws of Nature and in part incomplete human information about Nature—all scrambled up together by Bohr into am omelette that nobody has seen how to unscramble. Yet we think that the unscrambling is a prerequisite for any further advance in basic physical theory and we want to speculate on the proper tools to do this.

We suggest that the proper tool for incorporating human information into science is simply probability theory—not the currently taught "random variable" kind, but the original "logical inference" kind of James Bernoulli and Laplace. For historical reasons explained elsewhere (Jaynes 1986), this is often called "Bayesian probability theory." When supplemented by the notion of information entropy, this becomes a mathematical tool for scientific reasoning of such power and versatility that we think it will require a century to explore its capabilities. But the preliminary development of this tool and testing it on simple problems is now fairly well in hand, as described below.

~569~

A job for the immediate future is to see whether, by proper choice of variables, Bohr's omelette can be seen as a kind of approximation to it. In the 1950s, Richard Feynman noted that some of the probabilities in quantum theory obey different rules (interference of path amplitudes) than do the classical probabilities. But more recently (Jaynes 1989a) we have found that the QM probabilities involved in the EPR scenario are strikingly similar to the Bayesian probabilities, often identical; we interpret Bohr's reply to EPR as a recognition of this. That is, Bohr's explanation of the EPR experiment is a fairly good statement of Bayesian inference. Therefore, the omelette does have some discernible structure of the kind that we would need in order to unscramble it.

Probability as the Logic of Science

For some 200 years a debate has been underway on the philosophical level over this issue: Is probability theory a "physical" theory of phenomena governed by "chance" or "randomness," or is it an extension of logic, showing how to reason in situations of incomplete information? For two generations, the former view has dominated science almost completely.

More specifically, the basic equations of probability theory are the product and sum rules: denoting by AB the proposition that "A and B are both true" and by $\overline{A}$ the proposition that "A is false," these are

$$P(AB\,|\,C) = P(A\,|\,BC)P(B\,|\,C) = P(B\,|\,AC)P(A\,|\,C) \quad (2)$$

and

$$P(A\,|\,B) + P(\overline{A}\,|\,B) = 1, \quad (3)$$

and the issue is: What do these equations mean? Are they rules for calculating frequencies of "random variables" or rules for conducting plausible inference (reasoning from incomplete information)? Does the conditional probability symbol $P(A\,|\,B)$ stand for the frequency with which A is true in some "random experiment" defined by B, or for the degree of plausibility, in a single instance, that A is true given that B is true? Do probabilities describe real properties of Nature or only human information about Nature?

The original view of James Bernoulli and Laplace was that probability theory is an extension of logic to the case where, because of incomplete information, deductive reasoning by the Aristotelian syllogisms is not available. It was sometimes called "the calculus of inductive reasoning." All of Laplace's great contributions to science were made with the help of probability theory interpreted in this way.

But, starting in the mid-nineteenth century, Laplace's viewpoint came under violent attack from Leslie Ellis, John Venn, George Boole, R. von Mises, R. A. Fisher, M. G. Kendall, W. Feller, J. Neyman, and others. Their objection was to his philosophy; none of these critics was able to show that Laplace's methods (application of eqs. (2) and (3) as a form of logic) contained any inconsistency or led to any unsatisfactory results. Whenever they seemed to find such a case, closer examination always showed that they had only misunderstood and misapplied Laplace's methods.

Nevertheless, this school of thought was so aggressive that it has dominated the field almost totally, so that virtually all probability textbooks in current use are written from a viewpoint which rejects Laplace's interpretations and tries to deny us the use of his methods. Almost the only exceptions are found in the works of Harold Jeffreys (1939) and Arnold Zellner (1971), which recognize the merit of Laplace's viewpoint and apply it, with the same kind of good results that Laplace found, in more sophisticated current problems. We have written two short histories of these matters (Jaynes 1978, 1986), engaged in a polemical debate on them (Jaynes 1976), and are trying to finish a two-volume treatise on the subject, entitled *Probability Theory—The Logic of Science.* [1]

Denunciations of the "subjectivity" of Laplace, Jeffreys, and the writer for using probability to represent human information, and even more of the "subjectivity" of entropy based on such probabilities, often reach hysterical proportions; it is very hard to understand why so much emotional fervor should be aroused by these questions. Those who engage in these attacks are only making a public display of their own ignorance; it is apparent that their tactics amount to mere chanting of ideological slogans, while simply ignoring the relevant, demonstrable technical facts.

But the failure of our critics to find inconsistencies does not prove that our methods have any positive value for science. Are there any new useful results to be had from using probability

---

[1] *Eds.*: This work was published after Jaynes's death with the editorial assistance of G. L. Bretthorst. Jaynes, E. T. *Probability Theory: The Logic of Science*, edited by Bretthorst, G. L. 2003. Cambridge, UK: Cambridge University Press.

theory as logic? Some are reported in the proceedings volumes of the annual (since 1981) MAXENT workshops, particularly the one in Cambridge, England, in August 1988 wherein a generalized second law of thermodynamics is used in what we think is the first quantitative application of the second law in biology (Jaynes 1989a). But, unfortunately, most of the problems solvable by pencil-and-paper methods were too trivial to put this issue to a real test; although the results never conflicted with common sense, neither did they extend it very far beyond what common sense could see or what "random variable" probability theory could also derive.

Only recently, thanks to the computer, has it become feasible to solve real, non-trivial problems of reasoning from incomplete information, in which we use probability theory as a form of logic in situations where both intuition and "random variable" probability theory would be helpless. This has brought out the facts in a way that can no longer be obscured by arguments over philosophy. It is not easy to argue with a computer printout, which says to us: "Independently of all your philosophy, here are the facts about what this method actually gives when applied."

The MAXENT program developed by John Skilling, Steve Gull, and their colleagues at Cambridge University, England, can maximize entropy numerically in a space of $1,000,000$ dimensions, subject to $2,000$ simultaneous constraints. The "Bayesian" data-analysis program developed by G. L. Bretthorst (1988) at Washington University, St. Louis, can eliminate a hundred uninteresting parameters and give the simultaneous best estimates of twenty interesting ones and their accuracy, or it can take into account all the parameters in a set of possible theories or "models" and give us the relative probabilities of the theories in the light of the data. It was interesting, although to us not surprising, to find that this leads automatically to a quantitative statement of Occam's Razor: prefer the simpler theory unless the other gives a significantly better fit to the data.

Many computer printouts have now been made at Cambridge University of image reconstructions in optics and radio astronomy, and at Washington University in analyses of economic, geophysical, and nuclear magnetic resonance data. The results were astonishing to all of us; they could never have been found, or guessed, by hand methods.

In particular, the Bretthorst programs (Bretthorst et al. 1988; Bretthorst, Kotyk, and Ackerman 1989; Bretthorst and Ray Smith 1989) extract far more information from nuclear magnetic resonance data (where the ideal sinusoidal signals are corrupted by decay) than could the previously used Fourier transform methods. No longer does decay broaden the spectrum and obscure the information about oscillation frequencies; the result is an order-of-magnitude-better resolution.

Less spectacular numerically, but equally important in principle, they yield fundamental improvements in extracting information from economic time series when the data are corrupted by trend and seasonality; no longer do these obscure the information that we are trying to extract from the data. Conventional "random variable" probability theory lacks the technical means to eliminate nuisance parameters in this way, because it lacks the concept of "probability of a hypothesis."

In other words, there is no need to shout: it is now a very well-demonstrated fact that, after all criticisms of its underlying philosophy, probability theory interpreted and used as the logic of human inference does rather well in dealing with problems of scientific reasoning—just as James Bernoulli and Laplace thought it would, back in the eighteenth century.

Our probabilities and the entropies based on them are indeed "subjective" in the sense that they represent human information; if they did not, they could not serve their purpose. But they are completely "objective" in the sense that they are determined *by the information specified,* independently of anyone's personality, opinions, or hopes. It is "objectivity" in this sense that we need if information is ever to be a sound basis for new theoretical developments in science.

How Would Quantum Theory Be Different?

The aforementioned successful applications of probability theory as logic were concerned with data processing, while the original maximum entropy applications were in statistical mechanics, where they reproduced in a few lines and then generalized the results of Gibbs. In these applications, probability theory represented the process of reasoning from incomplete information. There is no claim that its predictions must be "right," only that they are the best that can be made from the information we have. (That is, after all, the most that any science can pretend to do; yet some complain when cherished illusions are replaced by honest recognition of the facts.)

We would like to see quantum theory in a similar way; since a pure state ψ does not contain enough information to predict all experimental results, we would like to see QM as the process of making the best predictions possible from the partial information that we have when we know ψ. If we could either succeed in this or prove that it is impossible, we would know far more about the basis of our present theory and about future possibilities for acquiring more information than we do today.

Einstein wanted to do something very similar, but he offered only criticisms rather than constructive suggestions.

What undoubtedly deterred both Einstein and Schrödinger is this: one sees quickly that the situation is more subtle than merely keeping the old mathematics and reinterpreting it. That is, we cannot merely proclaim that all the probabilities calculated within a QM pure state ψ according to the standard rules of our textbooks are now to be interpreted as expressions of human ignorance of the true physical state. The results depend on the representation in a way that makes this naive approach impossible.

~575~

For example, if we expand ψ in the energy representation $\psi = \sum a_n(t) u_n(x)$, the physical situation cannot be described merely as "the system may be in state $u_1(x)$ with probability $p_1 = |a_1|^2$, or it may be in state $u_2(x)$ with probability $p_2 = |a_2|^2$, and we do not know which of these is the true state." This would suffice to give, using classical probability theory, the QM predictions of quantities that are diagonal in the $\{u_n\}$ representation, but the relative phases of the amplitudes a_n have a definite physical meaning that would be lost by that approach.

Even though they have no effect on probabilities p_n in the energy representation, these phases will have a large effect on probabilities in some other representation. They affect the predicted values of quantities that are not diagonal in the $\{u_n\}$ representation, in a way that is necessary for agreement with experiment. For example, the relative phases of degenerate energy states of an atom determine the polarization of its resonance radiation, which is an experimental fact; so there has to be something physically real in them.

In other words, we cannot say merely that the atom is "in" state u_1 or "in" state u_2 as if they were mutually exclusive possibilities and it is only we who are ignorant of which is the true one; in some sense, it must be in both simultaneously or, as Pauli would say, the atom itself does not know what energy state it is in. This is the conceptually disturbing, but experimentally required, function of the superposition principle.

But notice that there is nothing conceptually disturbing in the statement that a vibrating bell is in a linear combination of two vibration modes with a definite relative phase; we just interpret the mode (amplitudes)^2 as *energies*, not *probabilities*. So it is the way we look at quantum theory, trying to interpret its symbols directly as probabilities, that is causing the difficulty.

If this seems at first to be an obstacle to our purpose, it is also our real opportunity, because it shows that the probabilities which we seek and which express the incompleteness of the information in a pure state in terms of a set of mutually exclusive possibilities (call it an "ensemble" if you like) cannot be the usual things called "probability" in the QM textbooks. The human information must be represented in a deeper "hypothesis space" which contains the phases as well as the amplitudes.

To realize this is to throw off a whole legacy of supposed difficulties from the past; the non-classical behavior of QM probabilities pointed out by Feynman ceases to bother us because the quantities exhibiting that behavior will not be interpreted as probabilities in the new hypothesis space. Likewise, the Bell inequality arguments are seen to have very little relevance to our problem, for he was hung up on the difficulty of getting the standard QM probabilities out of a causal theory. But if they are not the basic probabilities after all, the failure of a causal theory to reproduce them *as* probabilities might seem rather a merit than a defect. So the clouds begin to lift, just a bit.

This is not an auspicious time to be making public announcements of startling, revolutionary new scientific discoveries; so it is rather a relief that we have none to announce. To exhibit the variables of that deeper hypothesis space explicitly is a job for the future; in the meantime we can do a little job of housecleaning that is, in any event, a prerequisite for it. We cannot hope to get our probability connections right until we get some basic points of physics right.

The first difficulty we encounter upon any suggestion that probabilities in quantum theory might represent human information is the barrage of criticism from those who believe that dispersions $(\Delta F)^2 = \langle F^2 \rangle - \langle F \rangle^2$ represent experimentally observable "quantum fluctuations" in F. Some even claim that these fluctuations are real physical events that take place constantly whether or not any measurement is being made (although, of course, that does violence to Bohr's position). At the 1966 Rochester Coherence Conference, Roy Glauber assured us that vacuum fluctuations are "very real things" and that any attempts to dispense with EM field quantization are therefore doomed to failure. It can be reported that he was widely and enthusiastically believed.

Now in basic probability theory, ΔF represents fundamentally the accuracy with which we are able to *predict* the value of F. This does not deny that it may also be the variability seen in repeated measurements of F, but the point is that they need not be the same. To suppose that they *must* be the same is to commit an egregious form of the Mind Projection Fallacy; the fact that our information is able to determine F only to five percent accuracy is not enough to make it fluctuate by five percent! However, it is almost right to say that, given such information, any observed fluctuations are unlikely to be greater than five percent.

Let us analyze in depth the single example of EM field fluctuations and show that (1) the experimental facts do not require vacuum fluctuations to be real events after all; (2) Bayesian probability at this point not only is consistent with the experimental facts but offers us some striking advantages in clearing up past difficulties that have worried generations of physicists.

Is Zero-Point Energy Real?

For many years we have had a strange situation; on the one hand, "official quantum electrodynamics" has never taken the infinite zero-point (ZP) energy question seriously, apparently considering it only a formal detail like the infinite charge density in the original hole theory, which went away when the charge symmetry of the theory was made manifest in Schwinger's action principle formulation.

But the ZP problem has not gone away; on the other hand, as we have noted, there is a widespread belief that ZP fluctuations are real and necessary to account for all kinds of things, such as spontaneous emission, the Lamb shift, and the Casimir (1948) attraction effect. Steven Weinberg (1989) accepted the Casimir effect as demonstrating the reality of ZP energy, and worried about it in connection with cosmology. We know that Pauli also worried about this and did some calculations, but apparently never published them.

If one takes the ZP energy literally, one of the disturbing consequences is the gravitational field that it would produce. For example, if there is a ZP energy density W_{zp} in space, the Kepler ratio for a planet of mean distance R from the sun would be changed to

$$\frac{R^3}{T^2} = \frac{G}{4\pi^2} \left[M_{sun} + \frac{4\pi R^3}{3c^2} W_{zp} \right]. \qquad (4)$$

Numerical analysis of this shows that, in order to avoid conflict with the observed Kepler ratios of the outer planets, the upper frequency cutoff for the ZP energy would have to be taken no higher than optical frequencies.

But attempts to account for the Lamb shift by ZP fluctuations would require a cutoff thousands of times higher, at the Compton wavelength. The gravitational field from that energy density would not just perturb the Kepler ratio; it would completely disrupt the solar system as we know it.

-579-

The difficulty would disappear if one could show that the aforementioned effects have a different cause, and ZP field energy is not needed to account for any experimental facts. Let us try first with the simplest effect, spontaneous emission. The hypothesized ZP energy density in a frequency band $\Delta\omega$ is

$$W_{zp} = \rho_{zp}(\omega)\Delta\omega = \left(\frac{1}{2}\hbar\omega\right)\left(\frac{\omega^2\Delta\omega}{\pi^2 c^3}\right) \text{ ergs/cm}^3. \quad (5)$$

Then an atom decaying at a rate determined by the Einstein A-coefficient

$$A = \frac{4\mu^2\omega_0^3}{3\hbar c^3}, \quad (6)$$

where μ is the dipole moment matrix element for the transition, sees this over an effective bandwidth of

$$\Delta\omega = \frac{\int I(\omega)\,d\omega}{I(\omega_0)} = \frac{\pi A}{2}, \quad (7)$$

where $I(\omega)$ is the Lorentzian spectral density

$$I(\omega) \propto \frac{1}{(\omega - \omega_0)^2 + (A/2)^2}. \quad (8)$$

The effective energy density in one field component, say E_z, is then

$$(W_{zp})_{eff} = \frac{1}{6}\rho_{zp}(\omega)\Delta\omega = \frac{1}{18\pi}\mu^2\left(\frac{\omega}{c}\right)^6 \text{ ergs/cm}^3, \quad (9)$$

and it seems curious that Planck's constant has cancelled out. This indicates the magnitude of the electric field that a radiating atom sees according to the ZP theory.

On the other hand, the classical radiation reaction field generated by a dipole of moment μ,

$$E_{RR} = \frac{2}{3c^3} \frac{d^3 \mu}{dt^3} = \frac{2\omega^3}{3c^3} \mu, \qquad (10)$$

has energy density

$$W_{RR} = \frac{E_{RR}^2}{8\pi} = \frac{1}{18\pi} \mu^2 \left(\frac{\omega}{c}\right)^6 \text{ ergs/cm}^3. \qquad (11)$$

But equations (9) and (11) are identical! A radiating atom is indeed interacting with an electric field of just the magnitude predicted by the ZP calculation, but this is the atom's own radiation reaction field.

Now we can see that this needed field is generated by the radiating atom, automatically but in a more economical way: only where it is needed, when it is needed, and in the frequency band needed. Spontaneous emission does not require an infinite energy density throughout all space. Surely, this is a potentially far more satisfactory way of looking at the mechanism of spontaneous emission (if we can clear up some details about the dynamics of the process).

But then someone will point immediately to the Lamb shift; does this not prove the reality of the ZP energy? Indeed, Schwinger (1948, 1949) and Weisskopf (1949) stated explicitly that ZP field fluctuations are the physical cause of the Lamb shift, and Welton (1948) gave an elementary "classical" derivation of the effect from this premise.

Even Niels Bohr concurred. To the best of our knowledge, the closest he ever came to making an ontological statement was uttered while perhaps thrown momentarily off guard under the influence of Schwinger's famous eight-hour lecture at the

1948 Pocono conference. As recorded in John Wheeler's notes on that meeting, Bohr says: "It was a mistake in the older days to be discontented with field and charge fluctuations. They are necessary for the physical interpretation."

Dyson (1953) also concurred, picturing the quantized field as something akin to hydrodynamic flow with superposed random turbulence, and he wrote: "The Lamb–Retherford experiment is the strongest evidence we have for believing that our picture of the quantum field is correct in detail." Then in 1961 Feynman suggested that it should be possible to calculate the Lamb shift from the change in total ZP energy in space due to the presence of a hydrogen atom in the 2s state, and E. A. Power (1966) gave the calculation demonstrating this in detail. How can we possibly resist such a weight of authority and factual evidence?

As it turns out, quite easily. The problem has been that these calculations have been done heretofore only in a quantum field theory context. Because of this, people jumped to the conclusion that they were quantum effects (i.e., effects of field quantization), without taking the trouble to check whether they were present also in classical theory. As a result, two generations of physicists have regarded the Lamb shift as a deep, mysterious quantum effect that ordinary people cannot hope to understand. So we are facing not so much a weight of authority and facts as a mass of accumulated folklore.

Since our aim now is only to explain the elementary physics of the situation rather than to give a full formal calculation, let us show that this radiative frequency shift effect was present already in classical theory, and that its cause lies simply in properties of the source field (eq. 1), having nothing to do with field fluctuations. In fact, by stating the problem in Hamiltonian form, we can solve it without committing ourselves to electromagnetic or acoustical fields. Thus the vibrations of a plucked guitar string are also damped and shifted by their coupling to the acoustical radiation field, according to the following equations.

The Lamb Shift in Classical Mechanics

Let there be n "field oscillators" with coordinates and momenta $\{q_i(t), p_i(t)\}$, and one "extra oscillator" $\{Q(t), P(t)\}$, a caricature of a decaying atom or plucked string; call it "the EO." It is coupled linearly to the field oscillators, leading to a total Hamiltonian

$$H = \frac{1}{2} \sum_{i=1}^{n} (p_i^2 + \omega_i^2 q_i^2) + \frac{1}{2}(P^2 + \Omega^2 Q^2) - \sum_i \alpha_i q_i Q. \quad (12)$$

The physical effects of coupling the EO to the field variables may be calculated in two "complementary" ways.

(I) Dynamic: How are the EO oscillations modified by the field coupling?

(II) Static: What is the new distribution of normal mode frequencies?

The new normal modes are the roots $\{\nu_i\}$ of the equation $\Omega^2 - \nu^2 = K(\nu)$, where $K(\nu)$ is the dispersion function

$$K(\nu) \equiv \sum_i \frac{\alpha_i^2}{\omega_i^2 - \nu^2} = \int_0^\infty K(t) e^{-st}\, dt, \quad s = i\nu. \quad (13)$$

Let us solve the problem first in the more familiar dynamical way. With initially quiescent field modes $q_i(0) = \dot{q}_i(0) = 0$, the decay of the extra oscillator is found to obey a Volterra equation:

$$\ddot{Q}(t) + \Omega^2 Q(t) = \int_0^t K(t-t')Q(t')\,dt'. \tag{14}$$

Thus $K(t)$ is a memory function and the integral in equation (14) is a source field. For arbitrary initial EO conditions $\{Q(0), \dot{Q}(0)\}$, the solution is

-583-

$$Q(t) = Q(0)\dot{G}(t) + \dot{Q}(0)G(t) \tag{15}$$

with the Green's function

$$G(t) = \frac{1}{2\pi} \int_{-\infty}^{\infty} \frac{e^{i\nu t}\,d\nu}{\Omega^2 - \nu^2 - K(\nu)}, \tag{16}$$

where the contour goes under the poles on the real axis. This is the exact decay solution for arbitrary field mode patterns.

In the limit of many field modes, this goes into a simpler form. There is a mode density function $\rho_0(\omega)$:

$$\sum_i (\quad) \to \int_0^{\infty} (\quad)\rho_0(\omega)\,d\omega. \tag{17}$$

Then from equation (13), $K(\nu)$ goes into a slowly varying function on the path of integration for equation (16),

$$K(\nu - i\epsilon) \to \int_0^{\infty} \frac{\alpha^2(\omega)\rho_0(\omega)\,d\omega}{\omega^2 - (\nu - i\epsilon)^2} \to -2\nu[\Delta(\nu) + i\Gamma(\nu)], \tag{18}$$

and upon neglecting some small terms, the resulting Green's function goes into

$$G(t) \to \exp(-\Gamma t)\frac{\sin(\Omega + \Delta)t}{(\Omega + \Delta)}, \tag{19}$$

where

$$\Gamma(\Omega) \equiv \frac{\pi\alpha^2(\Omega)\rho_0(\Omega)}{4\Omega^2} \tag{20}$$

and

$$\Delta(\Omega) \equiv \frac{1}{2\Omega} P \int_0^\infty \frac{\alpha^2(\omega)\rho_0(\omega)\,d\omega}{\Omega^2 - \omega^2} = \frac{1}{\pi} P \int_{-\infty}^\infty \frac{\Gamma(\omega)\,d\omega}{\Omega - \omega} \quad (21)$$

are the "spontaneous emission rate" and "radiative frequency shift" exhibited by the EO due to its coupling to the field modes. We note that $\Delta(\Omega)$ and $\Gamma(\omega)$ form a Hilbert transform pair (a Kramers–Kronig-type dispersion relation expressing causality). In this approximation, equation (15) becomes the standard exponentially damped solution of a linear differential equation with loss: $\ddot{Q} + 2\Gamma\dot{Q} + (\Omega + \Delta)^2 Q = 0$.

As a check, it is a simple homework problem to compare our damping factor Γ with the well-known Larmor radiation law, by inserting into the above formulas the free-space mode density function $\rho_0(\omega) = V\omega^3/(\pi^2 c^3)$ and the coupling coefficients α_i appropriate to an electric dipole of moment μ proportional to Q. We then find

$$\Gamma(\omega) = \left(\frac{\pi}{4\omega^2}\right) \cdot \left(\frac{\mu^2}{Q^2} \cdot \frac{4\pi\omega^2}{3V}\right) \cdot \left(\frac{V\omega^2}{\pi^2 c^3}\right) = \frac{\mu^2\omega^2}{3Q^2 c^3} \sec^{-1}, \quad (22)$$

and it is easily seen that for the average energy loss over a cycle this agrees exactly with the Larmor formula

$$P_{\text{rad}} = \frac{2e^2}{3c^3}(\ddot{x})^2 \quad (23)$$

for radiation from an accelerated particle. In turn, the correspondence between the Larmor radiation rate and the Einstein A-coefficient in equation (6) is well-known textbook material.

It is clear from this derivation that the spontaneous emission and the radiative frequency shift do not require field fluctuations, since we started with the explicit initial condition of a quiescent field: $q_i = \dot{q}_i = 0$. The damping and shifting *are due entirely to the source field reacting back on the source,* as expressed by the integral in equation (14).

Of course, although the frequency shift formula in equation (21) resembles the "Bethe logarithm" expression for the Lamb shift, we cannot compare them directly because our model is not a hydrogen atom; we have no s-states and p-states. But if we use values of α_i and Ω for an electron oscillating at optical frequencies and use a cutoff corresponding to the size of the hydrogen atom, we get shifts of the order of magnitude of the Lamb shift. A more elaborate calculation will be reported elsewhere.

But now this seems to raise another mystery; if field fluctuations are not the cause of the Lamb shift, then why did the aforementioned Welton and Power calculations succeed by invoking those fluctuations? We face here a very deep question about the meaning of "fluctuation–dissipation theorems." There is a curious mathematical isomorphism; throughout this century, starting with Einstein's relation between the diffusion coefficient and mobility $D = \delta x^2/(2t) = kT\mu$ and the Nyquist thermal noise formula for a resistor $\delta V^2 = 4kTR\Delta f$, theoreticians have been deriving a steady stream of relations connecting "stochastic" problems with dynamical problems.

Indeed, for every differential equation with a non-negative Green's function, there is an obvious stochastic problem which would have the same mathematical solution even though the problems are quite unrelated physically; but as Mark Kac (1956) showed, the mathematical correspondence between stochastic and dynamical problems is much deeper and more general than that.

These relations do not prove that the fluctuations are real; they show only that certain dissipative effects (i.e., disappearance of the extra oscillator energy into the field modes) are the same *as if* fluctuations were present. But then by the Hilbert transform connection noted, the corresponding reactive effects must also be the same as if fluctuations were present; the calculation of Welton (1948) shows how this comes about.

But this still leaves a mystery surrounding the Feynman–Power calculation, which obtains the Lamb shift from the change in total ZP energy in the space surrounding the hydrogen atom; let us explain how that can be.

Classical Subtraction Physics

Consider now the second, static method of calculating the effect of field coupling. One of the effects of the EO is to change the distribution of normal modes; the above "free space" mode density $\rho_0(\omega)$ is incremented to

$$\rho(\omega) = \rho_0(\omega) + \rho_1(\omega). \tag{24}$$

To calculate the mode density increment, we need to evaluate the limiting form of the dispersion function $K(\nu)$ more carefully than in equation (18).

From the Hamiltonian in equation (12), the normal modes are the roots $\{\nu_i\}$ of the dispersion equation

$$\Omega^2 - \nu^2 = K(\nu) = \sum_i \frac{\alpha_i^2}{\omega_i^2 - \nu^2}. \tag{25}$$

$K(\nu)$ resembles a tangent function, having poles at the free field mode frequencies ω_i and zeros close to midway between them. Suppose that the unperturbed frequency Ω of the EO lies in the cell $\omega_i < \Omega < \omega_{i+1}$. Then the field modes above Ω are raised by amounts $\delta\nu_k = \nu_k - \omega_k$ for $k = i + 1, i + 2, \ldots, n$. The field modes below Ω are lowered by $\delta\nu_k = \nu_{k-1} - \omega_k$ for $k = 1, 2, \ldots, i$; and one new normal mode ν_i appears in the same cell as Ω: $\omega_i < \nu_i < \omega_{i+1}$. The separation property (exactly one new mode ν_k lies between any two adjacent old modes ω_i) places a stringent limitation on the magnitude of any static mode shift $\delta\nu_k$.

Thus the original field modes ω_i are, so to speak, pushed aside by a kind of repulsion from the added frequency Ω, and one new mode is inserted into the gap thus created. If there are many field modes, the result is a slight increase $\rho_1(\nu)$ in mode density in the vicinity of Ω. To calculate it, note that if the field mode ω_i is shifted a very small amount to $\nu_k = \omega_i + \delta\nu$, and $\delta\nu$ varies with ω_i, then the mode density is changed to

$$\rho(\omega) = \rho_0(\omega) + \rho_1(\omega) = \rho_0(\omega)\left[1 - \frac{d}{d\omega}(\delta\nu) + \cdots\right]. \quad (26)$$

In the continuum limit, $\rho_0 \to \infty$ and $\delta\nu \to 0$; however, the increment $\rho_1(\omega)$ remains finite and, as we shall see, loaded with physical meaning.

We now approximate the dispersion function $K(\nu)$ more carefully. In equation (16) where $\text{Im}(\nu) < 0$, we could approximate it merely by the integral, since the local behavior (the infinitely fine-grained variation in $K(\nu)$ from one pole to the next) cancels out in the limit at any finite distance from the real axis. But now we need it exactly on the real axis, and those fine-grained local variations are essential, because they provide the separation property that limits the static mode shifts $\delta\nu$.

Consider the case where $\omega_i > \Omega$ and ν lies in the cell $\omega_i < \nu < \omega_{i+1}$. Then the modes are pushed up. If the old modes near ν are about uniformly spaced, for small n we have $\omega_{i+n} \simeq \omega_i + n/\rho_0(\omega)$; therefore

$$\omega_{i+n}^2 - \nu^2 \simeq \frac{2\nu}{\rho_0}(n - \rho_0\delta\nu), \quad (27)$$

and the sum of terms with poles near ν goes into

$$\sum_n \frac{\alpha_{i+n}^2 \rho_0(\nu)}{2\nu(n - \rho_0 \delta\nu)} \simeq -\frac{\pi\alpha^2(\nu)\rho_0(\nu)}{2\nu} \cot[\pi\rho_0(\nu)\delta\nu], \quad (28)$$

where we supposed the α_i to be slowly varying and recognized the Mittag–Leffler expansion $\pi \cot \pi x = \sum (x - n)^{-1}$. The contribution of poles far from ν can again be represented by an integral. Thus, on the real axis, the dispersion function goes, in the continuum limit, into

$$K(\nu) \simeq -\frac{\pi\alpha^2\rho_0}{2\nu} \cot[\pi\rho_0(\nu)\delta\nu] + P \int_0^\infty \frac{\alpha^2(\omega)\rho_0(\omega)\, d\omega}{\omega^2 - \nu^2}.$$

But in this we recognize our expressions for Γ and Δ in equations (20) and (21):

$$K(\nu) \simeq -2\Omega[\Delta + \Gamma \cot(\pi\rho_0\delta\nu)]. \quad (29)$$

As a check, note that if we continue $\delta\nu$ below the real axis, the cotangent goes into $\cot(-ix) \to +i$ and we recover the previous result in equation (18). Thus if we again assume a sharp resonance $(\Omega \simeq \nu)$ and write the dynamically shifted frequency as $\omega_0 = \Omega + \Delta$, the dispersion relation in equation (25) becomes a formula for the static mode shift $\delta\nu$,

$$\pi\rho_0(\nu)\delta\nu = \tan^{-1}\left(\frac{\Gamma}{\nu - \omega_0}\right), \quad (30)$$

and (26) then yields for the increment in mode density a Lorentzian function:

$$\rho_1(\nu)\, d\nu = \frac{1}{\pi} \frac{\Gamma\, d\nu}{(\nu - \omega_0)^2 + \Gamma^2}. \quad (31)$$

This is the spectrum of a damped oscillation,

$$\int_{-\infty}^\infty \rho_1(\nu)e^{i\nu t}\, d\nu = e^{i\omega_0 t}e^{-\Gamma|t|}, \quad (32)$$

with the same shift and width as we found in the dynamical calculation of equation (14).

As a check, note that the increment is normalized, $\int \rho_1 \, d\nu = 1$, as it should be since the "macroscopic" effect of the coupled EO is just to add more new modes to the system. Note also that the result in equation (31) depended on $K(\nu)$ going locally into a tangent function. If for any reason (e.g., highly non-uniform mode spacing or coupling constants, even in the limit) $K(\nu)$ does not go into a tangent function, we will not get a Lorentzian $\rho_1(\nu)$. This would signify perturbing objects in the field, or cavity walls that do not recede to infinity in the limit, so echoes from them remain.

But the connection given in equation (32) between the mode density increment and the decay law is quite general. It does not depend on the Lorentzian form of $\rho_1(\nu)$, on the particular equation of motion for Q, on whether we have one or many resonances Ω, or indeed on any property of the perturbing EO other than the linearity of its response.

To see this, imagine that all normal modes are shock excited simultaneously with arbitrary amplitudes $A(\nu)$. Then the response is a superposition of all modes:

$$\int A(\nu)[\rho_0(\nu) + \rho_1(\nu)]e^{i\nu t} \, d\nu. \qquad (33)$$

But since the first integral represents the response of the free field, the second must represent the "ringing" of whatever perturbing objects are present. If $A(\nu)$ is nearly constant in the small bandwidth occupied by a narrow peak in $\rho_1(\nu)$, the resonant ringing goes into the form of equation (32).

Therefore, every detail of the transient decay of the dynamical problem is, so to speak, "frozen into" the static mode density increment function $\rho_1(\nu)$ and can be extracted by taking the Fourier transform shown in equation (32). Thus a bell excited by a pulse of sound will ring out at each of its resonant frequencies, each separate resonance having a decay rate and radiative frequency shift determined by $\rho_1(\nu)$ in the vicinity of that resonance.

Then a hydrogen atom in the 2s state excited by a sharp electromagnetic pulse will "ring out" at the frequencies of all the absorption or emission lines that start from the 2s state, and information about all the rates of decay and all the radiative line shifts is contained in the $\rho_1(\nu)$ perturbation that the presence of that atom makes in the field mode density.

Thus Feynman's conjecture about the relation between the Lamb shift and the change in ZP energy of the field around that atom is now seen to correspond to a perfectly general relation that was present all the time in classical electromagnetic and acoustical theory and might have been found by Rayleigh, Helmholtz, Maxwell, Larmor, Lorentz, or Poincaré in the last century.

It remains to finish the Power-type calculation and show that simple classical calculations can also be done by the more glamorous quantum-mechanical methods of "subtraction physics" if one wishes to do so. Suppose we put the extra oscillator in place and then turn on its coupling to the field oscillators. Before the coupling is turned on, we have a background mode density $\rho_0(\omega)$ with a single sharp resonance mode density $\delta(\omega - \Omega)$ superimposed. Turning on the coupling spreads this out into $\rho_1(\omega)$, superimposed on the same background, and shifts its center frequency by just the radiative shift Δ. In view of the normalization of $\rho_1(\omega)$, we can write

$$\Delta = \int_0^\infty \omega \rho_1(\omega) \, d\omega - \Omega. \tag{34}$$

Suppose, then, that we had asked a different question: "What is the total frequency shift *in all modes* due to the coupling?" Before the coupling is turned on, the total frequency is a badly divergent expression

$$(\infty)_1 = \Omega + \int_0^\infty \omega \rho_0(\omega)\, d\omega, \qquad (35)$$

and afterward it is

$$(\infty)_2 = \int_0^\infty \omega[\rho_0(\omega) + \rho_1(\omega)]\, d\omega, \qquad (36)$$

which is no better. But then the total change in all mode frequencies due to the coupling is, from equation (34),

$$(\infty)_2 - (\infty)_1 = \Delta. \qquad (37)$$

To do our physics by subtraction of infinities is an awkward way of asking the line-shift question, but it leads to the same result. There is no longer much mystery about why Power could calculate the radiative shift in the dynamical problem by the change in total ZP energy; actually, he calculated the change in total *frequency* of all modes, which was equal to the dynamical shift even in classical mechanics.

But some will still hold out and point to the Casimir attraction effect, where one measures a definite *force* which is held to arise from the change in total ZP energy when one changes the separation of two parallel metal plates. How could we account for this if the ZP energy is not real? This problem is already discussed in the literature; Schwinger, DeRaad, and Milton (1978) derive it from Schwinger's source theory, in which there are no operator fields. One sees the effect, like the van der Waals attraction, as arising from correlations in the states of electrons in the two plates, through the intermediary of their source fields as in equation (1). It does not require ZP energy to reside throughout all space, any more than does the van der Waals force. Thus we need not worry about the effect of ZP energy on the Kepler ratio in equation (4) or the cosmological constant, after all.

Conclusion

We have explored only a small part of the issues that we have raised; however, it is the part that has seemed the greatest obstacle to a unified treatment of probability in quantum theory. Its resolution was just a matter of getting our physics straight; we have been fooled by a subtle mathematical correspondence between stochastic and dynamical phenomena into believing that the "objective realities" of vacuum fluctuations and ZP energy are experimental facts. With the realization that this is not the case, many puzzling difficulties disappear.

We then see the possibility of a future quantum theory in which the role of incomplete information is recognized: the dispersion $(\Delta F)^2 = \langle F^2 \rangle - \langle F \rangle^2$ represents fundamentally only the accuracy with which the theory is able to *predict* the value of F. This may or may not be also the variability in the *measured* values.

In particular, if we free ourselves from the delusion that probabilities are physically real things, then when ΔF is infinite, that does not mean that any physical quantity is infinite. It means only that the theory is completely unable to predict F. The only thing that is infinite is the uncertainty of the prediction. In our view, this represents the beginning of a far more satisfactory way of looking at quantum theory, in which the important research problems will appear entirely different than they do now. ⸎

REFERENCES

Allen, L., and J. H. Eberly. 1975. *Optical Resonance and Two-Level Atoms*. Chap. 7. New York, NY: J. Wiley and Sons.

Bretthorst, G. L. 1988. *Bayesian Spectrum Analysis and Parameter Estimation*. Lecture Notes in Statistics, vol. 48. New York, NY: Springer. doi:10.1007/978-1-4684-9399-3.

Bretthorst, G. L., C. Hung, D. A. D'Avignon, and J. H. Ackerman. 1988. "Bayesian Analysis of Time-Domain Magnetic Resonance Signals." *Journal of Magnetic Resonance (1969)* 79 (2): 369–76. doi:10.1016/0022-2364(88)90233-8.

Bretthorst, G. L., J. J. Kotyk, and J. H. Ackerman. 1989. "^{31}P NMR Bayesian Spectral Analysis of Rat Brain *in vivo*." *Magnetic Resonance in Medicine* 9 (2): 282–87. doi:10.1002/mrm. 1910090214.

Bretthorst, G. L., and C. Ray Smith. 1989. "Bayesian Analysis of Signals from Closely Spaced Objects." In *Infrared Systems and Components III*, edited by R. L. Caswell, 1050:93–104. San Francisco, CA: SPIE.

Casimir, H. G. B. 1948. "On the Attraction between Two Perfectly Conducting Plates." *Proc. K. Ned. Akad. Wet.* 51:793–95.

Dyson, F. J. 1953. "Field Theory." *Scientific American* 188 (4): 57–64. doi:10 . 1038 / scientificamerican0453-57.

Jaynes, E. T. 1976. "Confidence Intervals vs. Bayesian Intervals." In *Foundations of Probability Theory, Statistical Inference, and Statistical Theories of Science,* edited by W. L. Harper and C. A. Hooker, 175–257. Reprinted in part in Jaynes (1989b). Dordrecht, Netherlands: D. Reidel Publishing Company.

———. 1978. "Where Do We Stand on Maximum Entropy?" In *The Maximum Entropy Formalism,* edited by R. D. Levine and M. Tribus. Reprinted in Jaynes (1989b). Cambridge, MA: MIT Press.

———. 1986. "Bayesian Methods: General Background." In *Maximum Entropy and Bayesian Methods in Applied Statistics,* edited by J. H. Justice, 1–25. Cambridge, UK: Cambridge University Press.

———. 1989a. "Clearing Up Mysteries: The Original Goal." In *Maximum Entropy and Bayesian Methods,* edited by J. Skilling, 1–27. Dordrecht, Netherlands: Kluwer Academic Publishers.

———. 1989b. *Papers on Probability, Statistics, and Statistical Physics*. Second paperback. Edited by R. D. Rosenkrantz. Reprints of 13 papers dated 1957–1980. Dordrecht, Netherlands: Kluwer Academic Publishers.

Jeffreys, H. 1939. *Probability Theory*. Later editions 1948, 1961, and 1966. A wealth of beautiful applications showing in detail how to use probability theory as logic. Oxford, UK: Oxford University Press.

Kac, M. 1956. *Some Stochastic Problems in Physics and Mathematics*. Colloquium Lectures in Pure and Applied Science, no. 2. Dallas, TX: Magnolia Petroleum Company.

Milonni, P. W., J. R. Ackerhalt, and W. A. Smith. 1973. "Interpretation of Radiative Corrections in Spontaneous Emission." *Physical Review Letters* 31 (15): 958–60. doi:10.1103/physrevlett.31.958.

Power, E. A. 1966. "Zero-Point Energy and the Lamb Shift." *American Journal of Physics* 34 (6): 516–18. doi:10.1119/1.1973082.

Schwinger, J. 1948. "Quantum Electrodynamics. I. A Covariant Formulation." *Physical Review* 74 (10): 1439–61. doi:10.1103/physrev.74.1439.

––––––. 1949. "Quantum Electrodynamics. II. Vacuum Polarization and Self-Energy." *Physical Review* 75 (4): 651–79. doi:10.1103/physrev.75.651.

Schwinger, J., L. L. DeRaad, and K. A. Milton. 1978. "Casimir Effect in Dielectrics." *Annals of Physics* 115 (1): 1–23. doi:10.1016/0003-4916(78)90172-0.

Senitzky, I. R. 1973. "Radiation-Reaction and Vacuum-Field Effects in Heisenberg-Picture Quantum Electrodynamics." *Physical Review Letters* 31 (15): 955–58. doi:10 . 1103 / physrevlett.31.955.

Weinberg, S. 1989. "The Cosmological Constant Problem." *Reviews of Modern Physics* 61 (1): 1–23. doi:10.1103/revmodphys.61.1.

Weisskopf, V. F. 1949. "Recent Developments in the Theory of the Electron." *Reviews of Modern Physics* 21 (2): 305–15. doi:10.1103/revmodphys.21.305.

Welton, T. A. 1948. "Some Observable Effects of the Quantum-Mechanical Fluctuations of the Electromagnetic Field." *Physical Review* 74 (9): 1157–67. doi:10.1103/physrev.74.1157.

Wigner, E. P. 1974. "Reminiscences on Quantum Theory." Colloquium talk at Washington University, St. Louis, March 27, 1974.

Zellner, A. 1971. *An Introduction to Bayesian Inference in Econometrics*. Reprinted by R. Krieger Publishing Company, Malabar, FL, 1987. The principles of Bayesian inference apply equally well in all fields, and all scientists can profit from this work. New York, NY: J. Wiley & Sons, Inc.

6 ᴺ

QUANTUM MEASUREMENTS AND ENTROPY

H. D. Zeh, Universität Heidelberg

Measurement-like quantum processes may lower the entropy or "create" unoccupied entropy capacity required for a thermodynamical arrow of time. The situation is also discussed in the Everett interpretation (where there is no collapse of the wave function) and for quantum gravity of a closed universe (where the wave function does not depend on a time parameter).

Introduction

Measurements in general are performed in order to increase information about physical systems. This information, if appropriate, may in principle be used for a reduction of their thermodynamical entropies—as we know from the thought construction of Maxwell's demon.

As we have been taught by authors including Smoluchowski, Szilard, Brillouin, and Gabor, one thereby has to invest *at least* the equivalent measure of information (therefore also called "negentropy") about a physical system in order to reduce its entropy by a certain amount. This is either required by the second law (*if* it is applicable for this purpose), or it can be derived within classical statistical mechanics by using

a. *determinism* and

b. the assumption that perturbations from outside may be treated stochastically in the forward direction of time (condition of "no conspiracy").

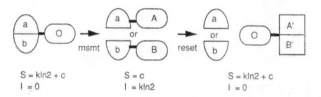

$S = k\ln 2 + c$ $S = c$ $S = k\ln 2 + c$
$I = 0$ $I = k\ln 2$ $I = 0$

Figure 1. Ensemble entropy and information in the deterministic description of a measurement and the subsequent reset.

The total *ensemble entropy* may then never decrease, and one can use diagrams such as figure 1 to represent sets of states (a, b) for the system being measured, the measurement and registration device (O, A, B), and the environment (A', B') which is required for the subsequent reset of the apparatus (Bennett 1973).

In statistical arguments of this kind, no concepts from phenomenological thermodynamics have to be used. Statistical counting is a more fundamental notion than energy conservation or temperature if the concept of deterministically evolving microscopic states is assumed to apply. The price to be paid for this advantage is the problem arising from the fact (much discussed at this conference) that the statistical ensemble entropy is not uniquely related to thermodynamical entropy.

This problem is even more important in the quantum-mechanical description. In quantum theory the statistical entropy is successfully calculated from the density matrix (regardless of the latter's interpretation). This density matrix changes *non-unitarily* (i.e., the state vectors diagonalizing it change *indeterministically*) in a measurement process (a situation usually referred to as the *collapse* or *reduction* of the wave function). So, for example, Pauli concluded that "the appearance of a certain result in a measurement is then a creation *outside the laws of nature.*" This may be a matter of definition—but the state vector (as it is used to describe an actual physical situation) *is* affected by the collapse, and so is the entropy calculated from it or from the density matrix!

Must this deviation from the deterministic Schrödinger equation now lead to a violation of the second law, as discussed in a beautiful way at this conference by Peres? In particular, can Maxwell's demon possibly return through the quantum back door?

In general, this is merely a question of principle. The amount of entropy corresponding to the information gain is extremely small compared to the thermodynamical entropy produced during the measurement (of the order of Landauer/Bennett's $k \ln 2$ discussed in computing). However, it will be argued that its effect may have been important during the early stages of the universe. In fact, it may even have been essential for the origin of Nature's arrow of time, which is based on an initially low value of entropy (Zeh 1989).

~599~

Questions of principle can only be answered within models. The model used for this discussion will be a universal quantum theory—either in its conventional form with a collapsing wave function (thus, according to M. Gell-Mann, making "concessions" to the traditional point of view) or by using some variant of the Everett interpretation. Therefore, no classical concepts will be used on a fundamental level. They will instead be considered as derivable. For the same reason, none of the arguments used will in any way be based on the uncertainty relations (in any other sense than the Fourier theorem).

First Reminder: The Arrow of Time in Radiation

Solutions of a hyperbolic-type differential equation can be represented in different ways, depending on the boundary conditions. For example, the electromagnetic fields can be written as

$$F^{\mu\nu} = F^{\mu\nu}_{\text{ret}} + F^{\mu\nu}_{\text{in}} = F^{\mu\nu}_{\text{adv}} + F^{\mu\nu}_{\text{out}},$$

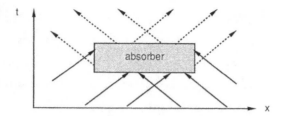

Figure 2. The time arrow of absorbers.

~600~ although in general one has

$$F^{\mu\nu} \approx F^{\mu\nu}_{\text{ret}} \text{ of "nearby" sources,}$$

where "nearby" may in astronomical situations include stars and galaxies. Eighty years ago, Einstein and Ritz (1909) required that this condition should hold by law of Nature, that is, exactly if considered for the whole universe. This assumption would eliminate the electromagnetic degrees of freedom and replace them by a retarded action at a distance. It corresponds to a cosmological initial condition ("Sommerfeld radiation condition") $F^{\mu\nu}_{\text{in}} = 0$.

A similar proposal was made seventy years later by Penrose (1979) for gravity instead of electrodynamics in terms of his *Weyl tensor hypothesis*. Both Ritz and Penrose expressed the expectation that their assumptions might then also explain the thermodynamical arrow of time.

The usual explanation of the electromagnetic arrow is that it is instead caused by the *thermodynamical arrow of absorbers* (see fig. 2): no field may leave an ideally absorbing region in the *forward* direction of time. (The same condition is required in the Wheeler–Feynman absorber theory *in addition* to their time-symmetric "absorber condition.")

The electrodynamical arrow of time can then easily be understood inside of closed laboratories possessing absorbing walls. In cosmology (where the specific boundary condition

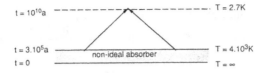

Figure 3. The radiation era as an early cosmic absorber.

is referred to as *Olbers' paradox*) the situation is slightly different. According to the big bang model there was a non-ideal (hot) absorber early in the universe (the radiation era; see fig. 3). Its thermal radiation has now cooled down to form the observed 2.7 K background radiation which is compatible with the boundary condition at wavelengths normally used in experiments. This early absorber hides the true $F_{in}^{\mu\nu}$ from view—although it is "transparent" for gravity.

However, it is important for many thermodynamical consequences that zero-mass fields possess a very large entropy capacity ("blank paper" in the language of information physics). This is true in particular for gravity because of the general attractivity and self-interaction that leads to the formation of black holes (Penrose 1981).

Second Reminder: The Thermodynamical Arrow in the Classical Statistical Description

In statistical mechanics the "irreversibility" of thermodynamical processes can be described in most general terms by means of Zwanzig's formalism of master equations. It consists of four essential steps:

1. Assume a unitary Liouville equation for ensembles $\rho(p, q, t)$,

$$\frac{i \partial \rho}{\partial t} = L\rho := i\{H, \rho\}.$$

It corresponds to the Hamiltonian *determinism* (conservation of probability) for the individual orbits.

Microscopic determinism can, however, only be a realistic assumption for the whole universe. This follows from a discussion of Borel (1924), who estimated the effect resulting from the displacement of a small mass on Sirius by a small distance on the microscopic state of a macroscopic gas here on earth. Within fractions of a second this state would thereupon become completely different from what it would have been otherwise. For this reason the "no-conspiracy condition" is essential.

2. Introduce an appropriate "concept of *relevance*" (or generalized coarse graining) by means of an idempotent operator $\mathbf{P}$:

$$\rho_{\text{rel}} = \mathbf{P}\rho \text{ with } \mathbf{P}^2 = \mathbf{P} \text{ and } \int \rho_{\text{rel}}\,dp\,dq = \int \rho\,dp\,dq.$$

This leads to a coupled dynamics for ρ_{rel} and $\rho_{\text{irrel}} = (1-\mathbf{P})\rho$ according to

$$\frac{i\,\partial\rho_{\text{rel}}}{\partial t} = \mathbf{P}L\rho_{\text{rel}} + \mathbf{P}L\rho_{\text{irrel}},$$

$$\frac{i\,\partial\rho_{\text{irrel}}}{\partial t} = (1-\mathbf{P})L\rho_{\text{rel}} + (1-\mathbf{P})L\rho_{\text{irrel}}.$$

Then formally solve the second equation for $\rho_{\text{irrel}}(t)$ with ρ_{rel} as an inhomogeneity (just as when calculating the electromagnetic field as a functional of the sources) and insert it into the first one to get the still exact (hence reversible) *pre-master equation* for ρ_{rel}:

$$\frac{i\,\partial\rho_{\text{rel}}}{\partial t} = \mathbf{P}L\rho_{\text{rel}}(t) + \mathbf{P}Le^{i(1-\mathbf{P})Lt}\rho_{\text{irrel}}(0)$$

$$- i\int_0^t G(\tau)\rho_{\text{rel}}(t-\tau)$$

$$\text{with } G(\tau) = \mathbf{P}Le^{-i(1-\mathbf{P})L\tau}(1-\mathbf{P})L\mathbf{P}.$$

The meanings of its three terms are indicated in figure 4. They correspond to a direct interaction for ρ_{rel}, an action of the initial $\rho_{\text{irrel}}(0)$ via the "irrelevant channel," and a retarded (if $t > 0$) non-Markovian action (corresponding to $F_{\text{ret}}^{\mu\nu}$), respectively. (This equation is valid for $t < 0$ as well.)

3. Eliminate the irrelevant degrees of freedom by assuming a *cosmological initial condition* $\rho_{\text{irrel}}(0) = 0$ (analogous to $F_{\text{in}}^{\mu\nu} = 0$ or Weyl$_{\text{in}} = 0$) and a large information capacity for ρ_{irrel} in order to approximately obtain the retarded *master equation* (which is Markovian after also assuming a slowly varying $\rho_{\text{rel}}(t)$)

$$\frac{\partial \rho_{\text{rel}}(t)}{\partial t} = -G_{\text{ret}}\rho_{\text{rel}}(t) \quad \text{with } G_{\text{ret}} = \int_0^\infty G(\tau)\, d\tau.$$

It represents an *"alternating dynamics"* (see fig. 5) consisting of the exact dynamics and the stepwise neglect of the irrelevant "information" which might otherwise flow back into the relevant channel. This is usually justified because of the initial condition and the large information capacity of the irrelevant degrees of freedom. The "doorway states" correspond to the radiation from the "nearby sources" in electrodynamics.

This master equation would become trivial in the limit $\Delta t \to 0$, which corresponds to what in quantum mechanics is called the watchdog effect or Zeno's quantum paradox (Joos 1984; Misra and Sudarshan 1977).

4. Define a *specific* concept of entropy (depending on **P**) by

$$S = -k \int \rho_{\text{rel}}(\ln \rho_{\text{rel}})\, dp\, dq$$

to obtain $dS/dt \geq 0$, that is, a monotonic loss of relevant information. In general, however, the equality $dS/dt \approx$

-603-

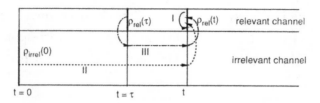

Figure 4. Information flow by means of Zwanzig's pre-master equation.

~604~

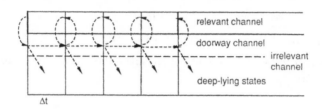

Figure 5. Alternating dynamics represented by the master equation.

0 would be overwhelmingly probable unless the further initial condition $S(t = 0) \ll S_{max}$ were to hold. This fact again possesses its analogue in electrodynamics: not all sources must be absorbers in order to prevent a trivial situation.

There exist *many* concepts of relevance (or "faces of entropy") suited for different purposes. Best known are Gibbs's coarse graining and Boltzmann's restriction to single-particle phase space (with the initial condition of absent particle correlations referred to as *molecular chaos*). They depend even further on the concept of particles used (elementary or compound, often changing during phase transitions or chemical reactions). Two others, P_{local} and P_{macro}, will be considered in more detail.

For most of the relevance concepts used, the condition $\rho_{irrel}(0) = 0$ does not appear very physical, since it refers to the *knowledge* described by the ensembles. Only some of them are effective on pure ("real") states, that is, they define a non-trivial entropy or "representative ensemble" as a function of *state*. All of them are based on a certain *observer-relatedness*: there is no *objective* reason for using ensembles or a concept of relevance. Also, Kolmogorov's entropy is based on a *relevant measure of distance*, while algorithms (used to define *algorithmic entropy*) are based on a choice of *relevant coordinates*. Hence, what we call chaos may merely be chaos *to us*!

Two Zwanzig projections will be of particular interest for illuminating the special character of quantum aspects. The first one is

$$\mathbf{P}_{\text{local}}\rho = \prod_i \rho_{\Delta V_i} \rightarrow S \approx \int s(\mathbf{r})\, d^3 r,$$

with *three-dimensional* volume elements ΔV_i containing many particles each. It is ineffective on pure classical states:

$$\mathbf{P}_{\text{local}}\delta^{6N} = \delta^{6N}.$$

The last statement is not true any more in quantum mechanics because of the existence of the fundamental quantum correlations which led to the violation of the Bell inequality.

The second projection of special interest is defined by

$$\mathbf{P}_{\text{macro}}\rho(p,q) = \text{const} =: \frac{p_\alpha}{V_\alpha} \text{ on } \alpha(p,q) = \text{const}$$

for "robust" (slowly changing and insensitive to perturbations) or "macroscopic" functions of state $\alpha(p,q)$. The dynamics within $\alpha = \text{const}$ may be assumed to be quasi-ergodic. The microscopic dynamics $(p(t), q(t))$ then induces a macroscopic dynamics $\alpha(t) := \alpha(p(t), q(t))$. This will again not remain true in quantum mechanics.

Under this second projection the entropy consists of two terms,

$$S[\mathbf{P}_{\text{macro}}\rho] = -k \sum p_\alpha \ln p_\alpha + \sum p_\alpha k \ln V_\alpha,$$

which represent the "lacking information" about the macroscopic quantities and the mean "physical entropy" $S(\alpha) = k \ln V_\alpha$

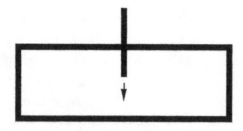

Figure 6. Deterministic transformation of "physical entropy" into "lacking information."

(Planck's "number of complexions"), respectively. This allows the *deterministic* transformation of physical entropy into "lacking information" (thereby conserving the ensemble entropy as in fig. 1). It is, in fact, part of Szilard's *gedanken* engine (fig. 6), where the transformation of entropy into lacking information renders the subdensities "robust." In its quantum version, this first part of the procedure may require the production of an additional, negligible but non-zero, amount of entropy in order to destroy the quantum correlations between the two partial volumes (Zurek 1986).

The Time Arrow of Quantum Measurements

The unitary quantum Liouville (von Neumann) equation,

$$\frac{i\,\partial\rho}{\partial t} = L\rho = [H, \rho],$$

corresponds again to the determinism (not to the unitarity) of the Schrödinger equation.

In quantum theory one often uses a specific relevance concept for the formulation of master equations. It is defined by the neglect of non-diagonal matrix elements,

$$\rho_{\text{rel}} = \mathbf{P}_{\text{diag}}\rho_{mn} = \rho_{mm}\delta_{mn},$$

with respect to a *given (relevant) basis.* Zwanzig's equation then becomes the van Hove equation (with an additional Born approximation the Pauli equation, or Fermi's Golden Rule after summing over final states). It has the form

$$\frac{d\rho_{mn}}{dt} = \sum_n A_{mn}(\rho_{nn} - \rho_{mm}),$$

with transition probabilities A_{mn} analogous to Boltzmann's *Stoßzahlansatz*. The meaning and validity of Zwanzig's approximation depends crucially on the choice of the "relevant basis." For example, it would become trivial $(A_{mn} = 0)$ in the exact energy basis.

In spite of its formal analogy to the classical theory, the quantum master equation describes the fundamental *quantum indeterminism*—not only an apparent indeterminism due to the lack of initial knowledge. For example, Pauli's equation is identical to the original probability interpretation of Born (1927) (which also introduced the Born approximation). It was to describe probabilities for new *wave functions* (not for classical particle positions), namely for the final states of the *quantum jumps* between Schrödinger's stationary eigenstates of the Hamiltonians of non-interacting local systems (which thus served as the dynamically "relevant basis"). Even these days the eigenstates of the second (and recently also of the "third") quantization are sometimes considered a "natural" and therefore fundamental basis of relevance for describing the collapse of the wave function as an *objective* process—although laser physicists, of course, know better.

Hence, the analogy to the classical theory is misleading. The reason is that the ensemble produced by the Zwanzig projection **P** from a pure state in general does not contain this state itself any more. According to the very foundation of the concept of the density matrix, it merely describes the probabilities for a collapse into the original state (or from it into another state).

In order to see this, the measurement process has to be considered in more detail. Following von Neumann's formulation one may write

$$\left(\sum c_n \phi_n \right) \Psi_0 \to \sum c_n \phi_n \Psi_n \to \phi_{n_0} \Psi_{n_0},$$

where the first step represents an appropriate interaction in accordance with the Schrödinger equation and the second step the collapse. I have left out an intermediate step leading to the ensemble of *potential* final states with their corresponding probabilities, since it describes only our ignorance of the outcome. The deterministic first step can again be realistic only if ψ represents the whole "rest of the universe," including the apparatus and the observer. This is the quantum analog of Borel's discussion of the extreme sensitivity of classical systems to their environments. Without the assumption of a wave function of the universe, no consistent theory would, therefore, be available at all—and no questions of principle could be answered.

The change of the ensemble entropy in the process of above is trivially given by

$$S_{\text{ensemble}} = 0 \to S_{\text{ensemble}} = 0 \to S_{\text{ensemble}} = 0,$$

whereas the corresponding local entropies are

$$S_{\text{local}} = 0 \to S_{\text{local}} \neq 0 \to S_{\text{local}} = 0,$$

since the intermediate state is non-local.

The fundamental *quantum indeterminism* is here represented by the fact that in the quantum formalism the initial ensemble entropy may vanish: *there is no ensemble of different initial states* (no "causes" or "sufficient reasons") for the different outcomes). This is in contrast to the classical situation represented by figure I. (The change in the ensemble entropy of any conjectured *hidden variables* would thus have to be compensated for in some unknown way during the measurement process.)

In the above description of a measurement, the "pointer position" Ψ_n must be a robust state Ψ_α in the sense mentioned before in order to form a genuine measurement. In this case the entropy changes according to $S_{\text{local}} \rightarrow S_{\text{physical}} = S(\alpha)$ during the collapse. The collapse is then part of the "objective" macroscopic dynamics $\alpha(t)$; in contrast to the classical description, no $\alpha(t)$ is induced by the Schrödinger equation.

On the other hand, Ψ_n is *not robust* for "measurement-like processes" without reading of the result, such as they occur in the continuous measurement by the environment which leads to *decoherence* (or rather to the *delocalization of phase relations*). The measurement-like processes thus cause the *local entropy to increase* by producing non-local correlations. (Cf. also Partovi's contribution to this conference.) For example, a small dust grain in intergalactic space produces entropy of the order of $S_{\text{local}} \approx k\ln(10^7)$ within 10^{-6} seconds (Joos and Zeh 1985). This is very small on a thermodynamical scale, although large compared to $k\ln 2$. The "irreversibility" of this process is caused by the assumption of a Sommerfeld radiation condition for the scattering—similar to Boltzmann's assumption of molecular chaos.

This result leads to the surprising consequence that classical mechanics (which in the absence of friction is usually considered to describe an exactly reversible dynamics) requires in quantum description the continuous action of irreversible processes for the classical properties to remain classical. The same is true for macroscopic computers: only strictly microscopic computers could in principle work reversibly.

Even the most effective information system (the genetic code) is macroscopic in this sense. Mother Nature may have her reasons, since this property seems to stabilize the information by means of the watchdog effect (Joos 1984). The effect also means that a (classically conceivable) *external* Laplacian demon (assumed to observe the world but not to react upon it) would have to decohere a quantum world.

The corresponding master equation of local relevance requires some initial condition like

$$(1 - \mathbf{P}_{local})\rho_{in} \approx 0 \quad \text{(no initial quantum correlations).}$$

In contrast to the classical theory, this condition is non-trivial even for a pure quantum state. It may, therefore, refer to "reality"—not merely to "our" knowledge about the initial state.

In principle, however, *the collapse may reduce the local entropy* according to the formalism of quantum theory! Although this is usually a small effect on a thermodynamical scale, it seems to occur in all phase transitions, situations of self-organization, etc.—whenever "virtual fluctuations" become macroscopic by the action of decoherence.

Lubkin (1987) has recently shown that this reduction of entropy cannot be used for the construction of a perpetuum mobile of the second kind—again, because of the required reset as studied by Bennett (1973). However, *there is no reset of the universe.* Therefore, consider the unique event of a phase transition of the vacuum in the early universe. It is most conveniently described as a transition between extrema of the Mexican hat potential in the form

$$|\Phi \equiv 0\rangle \to |\Phi \equiv \Phi_0 e^{i\phi}\rangle =: |\phi\rangle.$$

This process contains the collapse, since the Schrödinger equation with a symmetric Hamiltonian can only lead from the false vacuum $|\Phi \equiv 0\rangle$ to a symmetric superposition $\int |\phi\rangle \, d\phi$. Unless the false vacuum has the same energy expectation value as the physical vacuum, the state on the right-hand side must also contain excitations (which, in fact, contribute to the "measurement" of the value of ϕ characterizing a specific vacuum).

Except for the Casimir/Unruh correlations, the vacuum is a local state; that is, it can approximately be written as the same vacuum at every place, $|\phi\rangle \approx \prod_r |\phi\rangle_r$. This is not true for the symmetric superposition $\int |\phi\rangle \, d\phi$. Under the action of $\mathbf{P}_{\text{local}}$, this non-local state would instead lead to a mixed density matrix

$$\rho_{\mathbf{r}} \propto \sum_{-r} |\phi\rangle_r \langle\phi|_r.$$

Only the collapse leads then to a local zero-entropy state again, since it transforms a non-local state into a local state.

It appears suggestive that a similar mechanism created the degrees of freedom represented by the realistic zero-mass particles (photons and gravitons). This would correspond to the *creation* of a large unoccupied entropy capacity without deterministic "causes" (which would otherwise have to be counted by the previous values of the ensemble entropy as in fig. 1), or of "blank paper from nothing" by the symmetry-breaking power of the collapse.

These considerations have so far been based on the collapse as a process in time. However, on the one hand there is the (supposedly equivalent or even superior) Everett interpretation, which does not contain the collapse, and on the other hand there exists quantum gravity (necessarily to be taken into account in the whole wave function of the universe), which is stationary and does not contain any time parameter! The implications of these aspects for what happens to the entropy in quantum measurements will be discussed next.

The Everett Interpretation (No Collapse)

This interpretation is based on the sole validity of the deterministic Schrödinger equation with an essentially time-symmetric Hamiltonian. How, then, can it be equivalent to the collapse interpretation? How may the reduction of entropy induced by the collapse be understood in it?

The answer appears straightforward. The physical entropy $S(\alpha)$ is calculated as a function of the branches characterized by the macroscopic variables α (or "relative" to the state of the observer)—not as a functional of the total wave function. It is, hence, different in each branch Φ_α. However, the branching can be equivalent to the collapse only if the *arising* branches are afterwards dynamically independent from each other (or *robust*). But how can the time direction of this dynamically interpreted branching be compatible with the time-symmetric Schrödinger equation? Why are there only processes of the type $(\sum c_n \Phi_n)\Psi_0 \to \sum c_n \Phi_n \Psi_n$, with a robust state $\Psi_n = \Psi_\alpha$, and no inverse branchings of the type $\sum c_n \Phi_n \Psi_0 \to (\sum c_n \Phi_n)\Psi_n$?

Obviously, this requires an initial condition for the total Everett wave function, namely the condition of absence of matching "other" components $(n \neq n_0)$ *in the past*. Given local interactions only, this condition could well be of the type $\Psi_{in} = \Psi_{local} = \prod_i \Psi_{\Delta V_i}$ again, that is, the same as required for thermodynamics.

The quantum-mechanical arrow of time, therefore, appears in the Everett interpretation as an evolution from a *local initial state* into a more and more correlated state, that is, towards an increasing branching. It is thus formulated to be *fact-like* (as a specific property of the universal state), whereas the collapse would have to be considered *law-like*.

In the Everett interpretation, the deterministically evolving state vector may be considered as representing quantum "reality."

In this world model it is the observer whose identity changes indeterministically (subject to the branching), and so does the "relative state" of the universal wave function correlated to him. This is analogous to the process of cell division in normal space, which even in a Laplacian world would not determine one of the daughter cells as a unique successor. The "existence" of the other branches is concluded by hypothetically extrapolating the empirical laws (in this case the Schrödinger equation)—precisely as we conclude the existence of objects while not being observed, physical processes in the interior of the sun, and even events behind a spacetime horizon. Denying the Everett interpretation (or considering its other branches as mere "possibilities") is hence just another kind of solipsism!

This consideration emphasizes the observer-relatedness of the branching (and, therefore, of entropy). A candidate for its precise formulation may be the *Schmidt canonical* single-sum representation

$$\Psi(t) = \sum_n \sqrt{p_n(t)}\, \phi_n(t)\Phi_n(t)$$

with respect to any (local) observer system ϕ. It is unique (except for degeneracy) and therefore *defines* a "subjective" basis of relevance, although macroscopic properties contained in n seem to be objectivized by means of quantum correlations and the "irreversible" action of decoherence (Zeh 1971).

Quantum Gravity (No Time)

The dynamics of a quantum field theory that contains quantum gravity is described by the stationary Wheeler–DeWitt equation (Einstein–Schrödinger equation) or Hamiltonian constraint

$$H\Psi[^{(3)}G, \Phi(\mathbf{r})] = 0.$$

This equation does not allow one to impose an *initial condition* of low entropy in the usual way. How, then, can correlations such as those which are required to define the branching *evolve*?

The answer seems to be contained in the fact that the Wheeler–DeWitt Hamiltonian H is hyperbolic. For example, for Friedmann-type models with a massive quantum field (with its homogeneous part called Φ) one has

$$H = +\frac{\partial}{\partial\alpha^2} - \frac{\partial}{\partial\Phi^2} - \cdots + V(\alpha, \Phi, \ldots) =: +\frac{\partial}{\partial\alpha^2} + H_{\text{red}}^2,$$

where the dots refer to the higher multipoles of geometry and matter on the Friedmann sphere (Halliwell and Hawking 1985). This allows one to impose an initial condition with respect to the "intrinsic time" $\alpha = \ln a$, the logarithm of the Friedmann expansion parameter. The reduced dynamics H_{red} defines an intrinsic determinism, although not, in general, an intrinsic unitarity, since $V(\alpha, \Phi, \ldots)$ may be negative somewhere.

Because of the absence, in the wave function, of a term $\exp(i\omega t)$, there is no meaningful distinction between $\exp(+ika)$ and $\exp(-ika)$. (Halliwell—see his contribution to this volume— has presented arguments that these components decohere from another.) So the intrinsic big bang is identical to the intrinsic big crunch: they form one common, intrinsically initial, "big brunch."

On the other hand, because of the physical meaning of α, the potential V cannot be expected to be intrinsically "time"-symmetric under reversal of α. This asymmetry defines an intrinsic *dynamical* (law-like) arrow of time which is equal to that of the expansion of the universe.

This intrinsic dynamics gives rise to a paradox: whereas classical determinism may force its orbits to return in α (that is, the universe to recollapse), the intrinsic wave determinism allows one to exclude the "returning" parts of the corresponding wave packets by free choice of the "initial" conditions on a "time"-like hypersurface such as $\alpha = $ const. The paradox can be resolved by means of a "final" condition of *square integrability* of the wave function (hence $\Psi \Rightarrow 0$) for $\alpha \to \infty$. This condition is facilitated by the non-unitarity of the intrinsic dynamics for $H^2_{\mathrm{red}} < 0$. It forms half of a complete boundary condition which is different from Hartle and Hawking's but represents conventional quantum mechanics. Although there are no classically forbidden regions (since the kinetic energy is not positive definite), this "final" condition leads to a reflection of wave packets in "time" from the potential barrier which arises from a positive spatial curvature of the universe (Zeh 1988). The "initial" condition for the total wave function is then not completely free: the "returning" part of a wave packet must be present "initially," too.

- 615 -

A simple solvable example describing a situation of this kind is given by the normalizable eigenstates of the *indefinite anisotropic harmonic oscillator* with a rational ratio $\omega_\alpha : \omega_\Phi$,

$$H = -\frac{\partial^2}{\partial\alpha^2} + \frac{\partial^2}{\partial\Phi^2} + \omega_\alpha^2\alpha^2 - \omega_\Phi^2\Phi^2 - \text{ ground state energy.}$$

From its solutions one may construct coherent wave tubes which approximately define "orbits of causality" (see fig. 7) even when the actual wave function extends over the whole superspace. Similar behavior is found for other appropriate potentials, although the wave packets in general show dispersion towards the turning point in α (Kiefer 1988).

Corresponding wave functions in high-dimensional superspace show of course more complex behavior and may lead to an *increasing branching with increasing* α if an "initial" condition of lacking correlations holds for $\alpha \to -\infty$. If one, then, formally follows a turning classical orbit in mini- or midi-superspace, one should observe branching of the wave function for the microscopic variables on the expansion leg, but recombination (inverse branching) on the return leg. This point of view is, how-

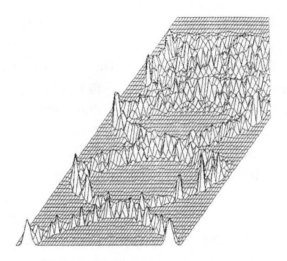

Figure 7. Wave tubes $\Psi(\alpha, \Phi)$ of the anisotropic indefinite harmonic oscillator (here plotted for $\alpha > 0$ and $\omega_\Phi : \omega_\alpha = 9 : 1$) define "orbits of causality." The symmetry between the two legs of the orbit is an artifact of this model. The intrinsic "relativistic" wave structure of the tubes is not resolved by the chosen grid size.

ever, merely a relict of the concept of classical orbits; the subjective arrow of time should in each leg be determined by the thermodynamical one. Closer to the turning point no clearly defined arrow can exist along the classical orbits, although the situation there seems to be very different from thermal equilibrium. The consequence of this (in the classical picture "double-ended") quantum version of the cosmic censorship postulate for the formation of black holes is not yet understood (Zeh 1983).

The time arrow of increasing quantum correlations leads, furthermore, to the very effective continuous measurement of the variables α and Φ by the higher multipoles (which act as an environment), and thereby to the *emergence* of a classical concept of time. In fact, these variables seem to form the "most classical" quantities of the universe (Kiefer 1987; Padmanabhan 1989).

Since the integrability condition seems to work well for $\alpha \to \infty$, could it not also be applicable to $\alpha \to -\infty$? This would complete the boundary condition in an entirely conventional manner. If it led to a unique solution, as already conjectured by DeWitt (1967), it would determine the wave function of the universe from its dynamics—a goal also aspired to by Hartle and Hawking's proposal.

This construction of the wave function of the universe would then have to *explain the cosmological initial condition of lacking correlations* from the "time"-asymmetry of the potential $V(\alpha, \ldots)$. For example, if superpositions of different spatial topologies can be neglected, an initial condition

$$\Psi \Rightarrow \Psi(\alpha \text{ only}) \to 0 \text{ for } \alpha \to -\infty$$

could simply be enforced by appropriate potential barriers for the multipole amplitudes. It would describe an initially "simple" (since unstructured) universe, although not the ground state of the higher multipoles on the Friedmann sphere. Any concept of ground or excited states could only be meaningful for them after they have entered their domain of adiabaticity.

This conceivable existence of a completely symmetric *pure* initial state (instead of a symmetric ensemble of very many states, the "real" member of which we were then unable to determine from this initial condition) is a specific consequence of the superposition principle, that is, of quantum mechanics. Before the "occurrence" of the first collapse or branching, the universe would then not contain any non-trivial degrees of freedom, or any potentiality of complexity.

This determination of the total wave function of the universe from its dynamics depends of course on the behavior of the realistic potential $V(\alpha, \Phi, \ldots)$ for $\alpha \to -\infty$. Since it refers to the Planck era, this procedure would require knowledge about a completely unified quantum field theory. Hopefully, this property of the potential may turn out to be a useful criterion for finding one! An appropriate potential for the higher modes would even be able to describe their effective cutoff at wavelengths of the order of the Planck length (useful for a finite renormalization) at *all* times. ↓

Acknowledgment

I wish to thank C. Kiefer and H. D. Conradi for their critical reading of the manuscript. Financial help from the Santa Fe Institute is acknowledged. This contribution was *not* supported by *Deutsche Forschungsgemeinschaft*.

REFERENCES

Bennett, C. H. 1973. "Logical Reversibility of Computation." *IBM Journal of Research and Development* 17 (6): 525–32. doi:10.1147/rd.176.0525.

Borel, E. 1924. *Le Hasard*. Paris, France: Alcan.

Born, M. 1927. "Das Adiabatenprinzip in der Quantenmechanik." *Zeitschrift für Physik* 40 (3–4): 167–92. doi:10.1007/bf01400360.

DeWitt, B. S. 1967. "Quantum Theory of Gravity. I. The Canonical Theory." *Physical Review* 160 (5): 1113–48. doi:10.1103/physrev.160.1113.

Einstein, A., and W. Ritz. 1909. "Zum gegenwärtigen Stand des Strahlungsproblems." *Physikalische Zeitschrift* 10:323–24.

Halliwell, J. J., and S. W. Hawking. 1985. "Origin of Structure in the Universe." *Physical Review D* 31 (8): 1777–91. doi:10.1103/physrevd.31.1777.

Joos, E. 1984. "Continuous Measurement: Watchdog Effect versus Golden Rule." *Physical Review D* 29 (8): 1626–33. doi:10.1103/physrevd.29.1626.

Joos, E., and H. D. Zeh. 1985. "The Emergence of Classical Properties through Interaction with the Environment." *Zeitschrift für Physik B Condensed Matter* 59 (2): 223–43. doi:10.1007/bf01725541.

Kiefer, C. 1987. "Continuous Measurement of Mini-Superspace Variables by Higher Multipoles." *Classical and Quantum Gravity* 4 (5): 1369–82. doi:10.1088/0264-9381/4/5/031.

————. 1988. "Wave Packets in Minisuperspace." *Physical Review D* 38 (6): 1761–72. doi:10.1103/physrevd.38.1761.

Lubkin, E. 1987. "Keeping the Entropy of Measurement: Szilard Revisited." *International Journal of Theoretical Physics* 26 (6): 523–35. doi:10.1007/bf00670091.

Misra, B., and E. C. G. Sudarshan. 1977. "The Zeno's Paradox in Quantum Theory." *Journal of Mathematical Physics* 18 (4): 756–63. doi:10.1063/1.523304.

Padmanabhan, T. 1989. "Decoherence in the Density Matrix Describing Quantum Three-Geometries and the Emergence of Classical Spacetime." *Physical Review D* 39 (10): 2924–32. doi:10.1103/physrevd.39.2924.

Penrose, R. 1979. "Singularities and Time-Asymmetry." In *General Relativity,* edited by S. W. Hawking and W. Israel. Cambridge, UK: Cambridge University Press.

————. 1981. "Time Asymmetry and Quantum Gravity." In *Quantum Gravity 2,* edited by C. J. Isham, R. Penrose, and D. W. Sciama. Oxford, UK: Clarendon Press.

Zeh, H. D. 1971. "On the Irreversibility of Time and Observation in Quantum Theory." In *Enrico Fermi School of Physics,* edited by B. d'Espagnat, vol. IL. New York, NY: Academic Press.

————. 1983. "Einstein Nonlocality, Space-Time Structure, and Thermodynamics." In *Old and New Questions in Physics, Cosmology, Philosophy, and Theoretical Biology,* edited by A. van der Merwe. New York, NY: Plenum.

————. 1988. "Time in Quantum Gravity." *Physics Letters A* 126 (5–6): 311–17. doi:10.1016/0375-9601(88)90842-0.

————. 1989. *The Physical Basis of the Direction of Time.* Heidelberg, Germany: Springer.

Zurek, W. H. 1986. "Maxwell's Demon, Szilard's Engine, and Quantum Measurements." In *Frontiers of Nonequilibrium Statistical Physics,* edited by G. T. Moore and M. T. Scully. New York, NY: Plenum.

QUANTUM THEORY &
MEASUREMENT

6ㅌ

QUANTUM MECHANICS IN THE
LIGHT OF QUANTUM COSMOLOGY

Murray Gell-Mann, California Institute of Technology
and James B. Hartle, University of California, Santa Barbara

~623~

We sketch a quantum-mechanical framework for the universe as a
whole. Within that framework we propose a program for describing
the ultimate origin in quantum cosmology of the "quasiclassical
domain" of familiar experience and for characterizing the process of
measurement. Predictions in quantum mechanics are made from
probabilities for sets of alternative histories. Probabilities (approximately
obeying the rules of probability theory) can be assigned only to sets of
histories that approximately decohere. Decoherence is defined and the
mechanism of decoherence is reviewed. Decoherence requires a sufficiently
coarse-grained description of alternative histories of the universe. A
quasiclassical domain consists of a branching set of alternative decohering
histories, described by a coarse graining that is, in an appropriate
sense, maximally refined consistent with decoherence, with individual
branches that exhibit a high level of classical correlation in time. We
pose the problem of making these notions precise and quantitative. A
quasiclassical domain is emergent in the universe as a consequence of
the initial condition and the action function of the elementary particles.
An important question is whether all the quasiclassical domains are
roughly equivalent or whether there are various essentially inequivalent
ones. A measurement is a correlation with variables in a quasiclassical
domain. An "observer" (or information gathering and utilizing system)
is a complex adaptive system that has evolved to exploit the relative
predictability of a quasiclassical domain, or rather a set of such
domains among which it cannot discriminate because of its own very
coarse graining. We suggest that resolution of many of the problems of
interpretation presented by quantum mechanics is to be accomplished not
by further scrutiny of the subject as it applies to reproducible laboratory
situations, but rather by an examination of alternative histories of the
universe, stemming from its initial condition, and a study of the problem
of quasiclassical domains.

I. Quantum Cosmology

If quantum mechanics is the underlying framework of the laws of physics, then there must be a description of the universe as a whole and everything in it in quantum-mechanical terms. In such a description, three forms of information are needed to make predictions about the universe. These are the action function of the elementary particles, the initial quantum state of the universe, and, since quantum mechanics is an inherently probabilistic theory, the information available about our specific history. These are sufficient for every prediction in science, and there are no predictions that do not, at a fundamental level, involve all three forms of information.

A unified theory of the dynamics of the basic fields has long been a goal of elementary particle physics and may now be within reach. The equally fundamental, equally necessary search for a theory of the initial state of the universe is the objective of the discipline of quantum cosmology. These may even be related goals; a single action function may describe both the Hamiltonian and the initial state.[1]

There has recently been much promising progress in the search for a theory of the quantum initial condition of the universe.[2] Such diverse observations as the large-scale homogeneity and isotropy of the universe, its approximate spatial flatness, the spectrum of density fluctuations from which the galaxies grew, the thermodynamic arrow of time, and the existence of classical spacetime may find a unified,

---

[1]As in the "no boundary" and the "tunneling from nothing" proposals where the wave function of the universe is constructed from the action by a Euclidean functional integral in the first case or by boundary conditions on the implied Wheeler–DeWitt equation in the second. See, e.g., Hartle and Hawking (1983) and Vilenkin (1986).

[2]For recent reviews see, e.g., Halliwell (1988b) and Hartle (1989a, 1990a). For a bibliography of papers on quantum cosmology, see Halliwell (1988a).

compressed explanation in a particular simple law of the initial condition.

The regularities exploited by the environmental sciences such as astronomy, geology, and biology must ultimately be traceable to the simplicity of the initial condition. Those regularities concern specific individual objects and not just reproducible situations involving identical particles, atoms, etc. The fact that the discovery of a bird in a forest or a fossil in a cliff or a coin in a ruin implies the likelihood of discovering another similar bird or fossil or coin cannot be derivable from the laws of elementary particle physics alone; it must involve correlations that stem from the initial condition.

The environmental sciences are not only strongly affected by the initial conditions but also heavily dependent on the outcomes of quantum-probabilistic events during the history of the universe. The statistical results of, say, proton–proton scattering in the laboratory are much less dependent on such outcomes. However, during the last few years there has been increasing speculation that, even in a unified fundamental theory, free of dimensionless parameters, some of the observable characteristics of the elementary particle system may be quantum-probabilistic, with a probability distribution that can depend on the initial condition.[3]

It is not our purpose in this article to review all these developments in quantum cosmology. Rather, we will discuss the implications of quantum cosmology for one of the subjects of this conference—the interpretation of quantum mechanics.

---

[3]As, for example, in recent discussions of the value of the cosmological constant; see, e.g., Hawking (1987), Coleman (1988), and Giddings and Strominger (1988).

II. Probability

Even apart from quantum mechanics, there is no certainty in this world; therefore physics deals in probabilities. In classical physics probabilities result from ignorance; in quantum mechanics they are fundamental as well. In the last analysis, even when treating ensembles statistically, we are concerned with the probabilities of particular events. We then deal in the probabilities of deviations from the expected behavior of the ensemble caused by fluctuations.

When the probabilities of particular events are sufficiently close to 0 or 1, we make a definite prediction. The criterion for "sufficiently close to 0 or 1" depends on the use to which the probabilities are put. Consider, for example, a prediction on the basis of present astronomical observations that the sun will come up tomorrow at 5:59 a.m. $\pm$ 1 minute. Of course, there is no certainty that the sun will come up at this time. There might have been a significant error in the astronomical observations or the subsequent calculations using them; there might be a non-classical fluctuation in the earth's rotation rate; or there might be a collision with a neutron star now racing across the galaxy at near light speed. The prediction is the same as estimating the probabilities of these alternatives as low. How low do they have to be before one sleeps peacefully tonight rather than anxiously awaiting the dawn? The probabilities predicted by the laws of physics and the statistics of errors are generally agreed to be low enough!

All predictions in science are, most honestly and most generally, the probabilistic predictions of the *time histories* of particular events in the universe. In cosmology we are necessarily concerned with probabilities for the single system that is the universe as a whole. Where the universe presents us effectively with an ensemble of identical subsystems, as in experimental situations common in physics and chemistry, the probabilities for the ensemble as a whole yield definite predictions for the statistics of identical observations. Thus, statistical probabilities can be derived, in appropriate situations, from probabilities for the universe as a whole (Finkelstein 1963; Hartle 1968; Graham 1973; Farhi, Goldstone, and Gutmann 1989).

Probabilities for histories need be assigned by physical theory only to the accuracy to which they are used. Thus, it is the same to us for all practical purposes whether physics claims the probability of the sun not coming up tomorrow is $10^{-10^{37}}$ or $10^{-10^{27}}$, as long as it is very small. We can therefore conveniently consider *approximate probabilities*, which need obey the rules of the probability calculus only up to some standard of accuracy sufficient for all practical purposes. In quantum mechanics, as we shall see, it is likely that only by this means can probabilities be assigned to interesting histories at all.

III. Historical Remarks

In quantum mechanics not every history can be assigned a probability. Nowhere is this more clearly illustrated than in the two-slit experiment (fig. 1). In the usual discussion, if we have not measured which slit the electron passed through on its way to being detected at the screen, then we are not permitted to assign probabilities to these alternative histories. It would be inconsistent to do so since the correct probability sum rules would not be satisfied. Because of interference, the probability of arriving at y is not the sum of the probabilities of arriving at y going through the upper and the lower slits:

$$p(y) \neq p_U(y) + p_L(y), \tag{1}$$

because

$$|\psi_L(y) + \psi_U(y)|^2 \neq |\psi_L(y)|^2 + |\psi_U(y)|^2. \tag{2}$$

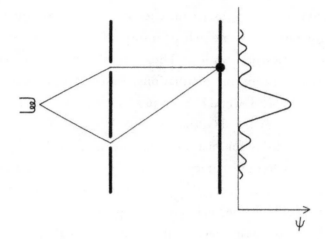

Figure 1. The two-slit experiment. An electron gun at the left emits an electron traveling toward a screen with two slits, its progress in space recapitulating its evolution in time. When precise detections are made of an ensemble of such electrons at the screen, it is not possible, because of interference, to assign a probability to the alternatives of whether an individual electron went through the upper slit or the lower slit. However, if the electron interacts with apparatus which measures which slit it passed through, then these alternatives decohere and probabilities can be assigned.

If we *have* measured which slit the electron went through, then the interference is destroyed, the sum rule is obeyed, and we *can* meaningfully assign probabilities to these alternative histories.

It is a general feature of quantum mechanics that one needs a rule to determine which histories can be assigned probabilities. The familiar rule of the "Copenhagen" interpretations described above is external to the framework of wave function and Schrödinger equation. Characteristically these interpretations, in one way or another, assumed as fundamental the existence of the classical domain we see all about us. Bohr spoke of phenomena that could be described in terms of classical language.[4] Landau and Lifshitz (1958) formulated quantum mechanics in terms of a separate classical physics. Heisenberg and others stressed the

---

[4]See the essays "The Unity of Knowledge" and "Atoms and Human Knowledge," reprinted in Bohr (1958).

central role of an external, essentially classical observer.[5] A measurement occurred through contact with this classical domain. Measurements determined what could be spoken about.

Such interpretations are inadequate for cosmology. In a theory of the whole thing there can be no fundamental division into observer and observed. Measurements and observers cannot be fundamental notions in a theory that seeks to discuss the early universe when neither existed. There is no reason in general for a classical domain to be fundamental or external in a basic formulation of quantum mechanics.

~ 629 ~

It was Everett who in 1957 first suggested how to generalize the Copenhagen framework so as to apply quantum mechanics to cosmology.[6] His idea was to take quantum mechanics seriously and apply it to the universe as a whole. He showed how an observer could be considered part of this system and how its activities—measuring, recording, and calculating probabilities—could be described in quantum mechanics.

Yet the Everett analysis was not complete. It did not adequately explain the origin of the classical domain or the meaning of the "branching" that replaced the notion of measurement. It was a theory of "many worlds" (what we would rather call "many histories"), but it did not sufficiently explain how these were defined or how they arose. Also, Everett's discussion suggests that a probability formula is somehow not needed in quantum mechanics, even though a "measure" is introduced that, in the end, amounts to the same thing.

---

[5]For clear statements of this point of view, see London and Bauer (1939) and Peierls (1985).

[6]The original paper is Everett (1957). The idea was developed by many, among them Wheeler (1957), DeWitt (1970), Geroch (1984), and Mukhanov (1985), and independently arrived at by others, e.g., Gell-Mann (1963) and Cooper and Van Vechten (1969). There is a useful collection of early papers on the subject in DeWitt and Graham (1973).

Here we shall briefly sketch a program aiming at a coherent formulation of quantum mechanics for science as a whole, including cosmology as well as the environmental sciences.[7] It is an attempt at extension, clarification, and completion of the Everett interpretation. It builds on many aspects of the post-Everett developments, especially the work of Zeh (1970), Zurek (1981, 1982), and Joos and Zeh (1985). In the discussion of history and at other points it is consistent with the insightful work (independent of ours) of Griffiths (1984) and Omnès (1988a, 1988b, 1988c). Our research is not complete, but we sketch, in this report on its status, how it might become so.

IV. Decoherent Sets of Histories

(A) A CAVEAT

We shall now describe the rules that specify which histories may be assigned probabilities and what these probabilities are. To keep the discussion manageable we make one important simplifying approximation: we neglect gross quantum variations in the structure of spacetime. This approximation, excellent for times later than 10^{-43} seconds after the beginning, permits us to use any of the familiar formulations of quantum mechanics with a preferred time. Since histories are our concern, we shall often use Feynman's sum-over-histories formulation of quantum mechanics with histories specified as functions of this time. Since the Hamiltonian formulation of quantum mechanics is in some ways more flexible, we shall use it also, with its apparatus of Hilbert space, states, Hamiltonian, and other operators. We shall indicate the equivalence between the two, always possible in this approximation.

---

[7]Some elements of which have been reported earlier; see Gell-Mann (1987).

The approximation of a fixed background spacetime breaks down in the early universe. There, a yet more fundamental sum-over histories framework of quantum mechanics may be necessary.[8] In such a framework the notions of state, operators, and Hamiltonian may be approximate features appropriate to the universe after the Planck era, for particular initial conditions that imply an approximately fixed background spacetime there. A discussion of quantum spacetime is essential for any detailed theory of the initial condition, but when, as here, this condition is not spelled out in detail and we are treating events after the Planck era, the familiar formulation of quantum mechanics is an adequate approximation.

The interpretation of quantum mechanics that we shall describe in connection with cosmology can, of course, also apply to any strictly closed subsystem of the universe provided its initial density matrix is known. However, strictly closed subsystems of any size are not easily realized in the universe. Even slight interactions, such as those of a planet with the cosmic background radiation, can be important for the quantum mechanics of a system, as we shall see. Further, it would be extraordinarily difficult to prepare precisely the initial density matrix of any sizeable system so as to get rid of the dependence on the density matrix of the universe. In fact, even those large systems that are approximately isolated today inherit many important features of their effective density matrix from the initial condition of the universe.

(B) HISTORIES

The three forms of information necessary for prediction in quantum cosmology are represented in the Heisenberg picture

---

[8]See, e.g., Hartle (1988a, 1988b, 1989b, 1990b, 1990c). For a concise discussion see Gell-Mann (1989).

as follows:[9] The quantum state of the universe is described by a density matrix ρ. Observables describing specific information are represented by operators $\mathcal{O}(t)$. For simplicity, but without loss of generality, we shall focus on non-"fuzzy", "yes–no" observables. These are represented in the Heisenberg picture by projection operators $P(t)$. The Hamiltonian, which is the remaining form of information, describes evolution by relating the operators corresponding to the same question at different times through

$$P(t) = e^{iHt/\hbar} P(0) e^{-iHt/\hbar}. \tag{3}$$

An exhaustive set of "yes–no" alternatives at one time is represented in the Heisenberg picture by *sets* of projection operators $(P_1^k(t), P_2^k(t), \ldots)$. In $P_\alpha^k(t)$, k labels the set, α the particular alternative, and t its time. A exhaustive set of exclusive alternatives satisfies

$$\sum_\alpha P_\alpha^k(t) = 1, \quad P_\alpha^k P_\beta^k = \delta_{\alpha\beta} P_\alpha^k. \tag{4}$$

For example, one such exhaustive set would specify whether a field at a point on a surface of constant t is in one or another of a set of ranges exhausting all possible values. The projections are simply the projections onto eigenstates of the field at that point with values in these ranges. We should emphasize that an exhaustive set of projections need not involve a *complete* set of variables for the universe (one-dimensional projections)—in fact, the projections we deal with as observers of the universe typically involve only an infinitesimal fraction of a complete set.

Sets of alternative histories consist of *time sequences* of exhaustive sets of alternatives. A *history* is a particular sequence of

---

[9]The utility of this Heisenberg picture formulation of quantum mechanics has been stressed by many authors, among them Wigner (1963), Aharonov, Bergmann, and Lebowitz (1964), Unruh (1986), and Gell-Mann (1987).

alternatives, abbreviated $[P_\alpha] = (P_{\alpha_1}^1(t_1), P_{\alpha_2}^2(t_2), \ldots, P_{\alpha_n}^n(t_n))$. A *completely fine-grained history* is specified by giving the values of a complete set of operators at all times. One history is a *coarse graining* of another if the set $[P_\alpha]$ of the first history consists of sums of the $[P_\alpha]$ of the second history. The inverse relation is fine graining. The completely coarse-grained history is one with no projections whatever, just the unit operator!

The reciprocal relationships of coarse and fine graining evidently constitute only a partial ordering of sets of alternative histories. Two arbitrary sets need not be related to each other by coarse/fine graining. The partial ordering is represented schematically in figure 2, where each point stands for a set of alternative histories.

Feynman's sum-over-histories formulation of quantum mechanics begins by specifying the amplitude for a completely fine-grained history in a particular basis of generalized coordinates $Q^i(t)$, say all fundamental field variables at all points in space. This amplitude is proportional to

$$\exp(iS[Q^i(t)]/\hbar), \qquad (5)$$

where S is the action functional that yields the Hamiltonian, H. When we employ this formulation of quantum mechanics, we shall introduce the simplification of ignoring fields with spins higher than zero, so as to avoid the complications of gauge groups and of fermion fields (for which it is inappropriate to discuss eigenstates of the field variables.) The operators $Q^i(t)$ are thus various scalar fields at different points of space.

Let us now specialize our discussion of histories to the generalized coordinate bases $Q^i(t)$ of the Feynman approach. Later we shall discuss the necessary generalization to the case of an arbitrary basis at each time t, utilizing quantum-mechanical transformation theory.

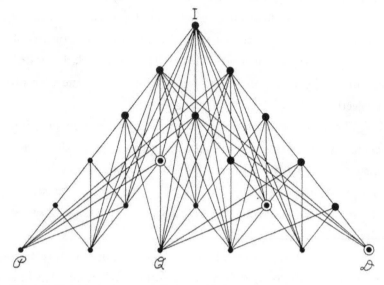

Figure 2. The schematic structure of the space of *sets* of possible histories for the universe. Each dot in this diagram represents an exhaustive set of alternative histories. Such sets, denoted by $\{[P_\alpha]\}$ in the text, correspond in the Heisenberg picture to time sequences $(P^1_{\alpha_1}(t_1), P^2_{\alpha_2}(t_2), \ldots, P^n_{\alpha_n}(t_n))$ of sets of projection operators, such that at each time t_k the alternatives α_k are an orthogonal and exhaustive set of possibilities for the universe. At the bottom of the diagram are the completely fine-grained sets of histories, each arising from taking projections onto eigenstates of a *complete set* of observables for the universe at *every time*. For example, the set Q is the set in which all field variables at all points of space are specified at every time. P might be the completely fine-grained set in which all field momenta are specified at each time. D might be a degenerate set of the kind discussed in the section "Maximal Sets of Decohering Histories" in which the *same* complete set of *operators* occurs at every time. But there are many other completely fine-grained sets of histories corresponding to all possible combinations of complete sets of observables that can be taken at every time.

The dots above the bottom row are coarse-grained sets of alternative histories. If two dots are connected by a path, the one above is a coarse graining of the one below—that is, the projections in the set above are *sums* of those in the set below. At the very top is the degenerate case in which complete sums are taken at every time, yielding no projections at all other than the unit operator! The space of sets of alternative histories is thus partially ordered by the operation of coarse graining.

The heavy dots denote the decoherent sets of alternative histories. Coarse grainings of decoherent sets remain decoherent. Maximal sets, the heavy dots surrounded by circles, are those decohering sets for which there is no finer-grained decoherent set.

Completely fine-grained histories in the coordinate basis cannot be assigned probabilities; only suitable coarse-grained histories can. There are at least three common types of coarse graining: (1) specifying observables not at all times, but only at some times; (2) specifying at any one time not a complete set of observables, but only some of them; (3) specifying for these observables not precise values, but only ranges of values. To illustrate all three, let us divide the Q^i up into variables x^i and X^i and consider only sets of ranges $\{\Delta_\alpha^k\}$ of x^i at times $t_k, k = 1, \ldots, n$. A set of alternatives at any one time consists of ranges Δ_α^k, which exhaust the possible values of x^i as α ranges over all integers. An individual history is specified by particular Δ_αs at the times $t_1, \ldots, t_n$. We write $[\Delta_\alpha] = (\Delta_{\alpha_1}^1, \ldots, \Delta_{\alpha_n}^n)$ for a particular history. A *set* of alternative histories is obtained by letting $\alpha_1, \ldots, \alpha_n$ range over all values.

Let us use the same notation $[\Delta_\alpha]$ for the most general history that is a coarse graining of the completely fine-grained history in the coordinate basis, specified by ranges of the Q^i at each time, including the possibility of full ranges at certain times, which eliminate those times from consideration.

(C) DECOHERING HISTORIES

The important theoretical construct for giving the rule that determines whether probabilities may be assigned to a given set of alternative histories, and what these probabilities are, is the decoherence functional $D[(\text{history})', (\text{history})]$. This is a complex functional on any pair of histories in the set. It is most transparently defined in the sum-over-histories framework for completely fine-grained history segments between an initial time

t_0 and a final time t_f, as follows:

$$D[Q'^i(t), Q^i(t)]$$
$$= \delta(Q_f'^i - Q_f^i) \exp\{i(S[Q'^i(t)] - S[Q^i(t)])/\hbar\}\rho(Q_0'^i, Q_0^i).$$
$$(6)$$

Here ρ is the initial density matrix of the universe in the Q^i representation, $Q_0'^i$ and Q_0^i are the initial values of the complete set of variables, and $Q_f'^i$ and Q_f^i are the final values. The decoherence functional for coarse-grained histories is obtained from equation (6) according to the principle of superposition by summing over all that is not specified by the coarse graining. Thus,

$$D([\Delta_{\alpha'}], [\Delta_\alpha])$$
$$= \int_{[\Delta_{\alpha'}]} \delta Q' \int_{[\Delta_\alpha]} \delta Q \, \delta(Q_f'^i - Q_f^i) \, e^{i\{(S[Q'^i]-S[Q^i])/\hbar\}} \rho(Q_0'^i, Q_0^i).$$
$$(7)$$

More precisely, the integral is as follows (fig. 3): It is over all histories $Q'^i(t)$ and $Q^i(t)$ that begin at $Q_0'^i$ and Q_0^i respectively, pass through the ranges $[\Delta_{\alpha'}]$ and $[\Delta_\alpha]$ respectively, and wind up at a common point Q_f^i at any time $t_f > t_n$. It is completed by integrating over $Q_0'^i$, Q_0^i, and Q_f^i.

The connection between coarse-grained histories and completely fine-grained ones is transparent in the sum-over-histories formulation of quantum mechanics.

However, the sum-over-histories formulation does not allow us to consider directly histories of the most general type. For the most general histories one needs to exploit directly the transformation theory of quantum mechanics, and for this the Heisenberg picture is convenient. In the Heisenberg picture D can be written as

$$D([P_{\alpha'}], [P_\alpha]) = \mathrm{Tr}[P_{\alpha_n}^n(t_n) \cdots P_{\alpha_1}^1(t_1)\rho P_{\alpha_1}^1(t_1) \cdots P_{\alpha_n}^n(t_n)].$$
$$(8)$$

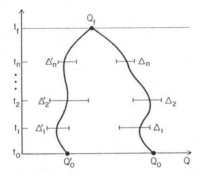

Figure 3. The sum-over-histories construction of the decoherence functional.

The projections in equation (8) are time ordered with the earliest on the inside. When the Ps are projections onto ranges Δ_α^k of values of the Qs, the expressions in equations (7) and (8) agree. From the cyclic property of the trace it follows that D is always diagonal in the final indices α_n and α'_n. (We assume throughout that the Ps are bounded operators in Hilbert space dealing, for example, with projections onto ranges of the Qs and not onto definite values of the Qs.) Decoherence is thus an interesting notion only for strings of Ps that involve more than one time. Decoherence is automatic for "histories" that consist of alternatives at but one time.

Progressive coarse graining may be seen in the sum-over-histories picture as summing over those parts of the fine-grained histories not specified in the coarse-grained one, according to the principle of superposition. In the Heisenberg picture, equation (8), the three common forms of coarse graining discussed above can be represented as follows: Summing on both sides of D over all Ps at a given time and using equation (3) eliminates those Ps completely. Summing over all possibilities for certain variables at one time amounts to factoring the Ps and eliminating one of the factors by summing over it. Summing over ranges of values of a given variable at a given time corresponds to replacing the Ps for the partial ranges by one for the total range. Thus, if $[\overline{P_\beta}]$ is a coarse graining of the set of histories $\{[P_\alpha]\}$, we write

$$D([\overline{P_{\beta'}}], [\overline{P_\beta}]) = \sum_{\substack{\text{all } P'_\alpha \\ \text{not fixed by } [\overline{P_{\beta'}}]}} \sum_{\substack{\text{all } P_\alpha \\ \text{not fixed by } [\overline{P_\beta}]}} D([P_{\alpha'}], [P_\alpha]).$$

(9)

In the most general case, we may think of the completely fine-grained limit as being obtained from the coordinate representation by arbitrary unitary transformations at all times. All histories can be obtained by coarse graining the various completely fine-grained ones, and coarse graining in its most general form involves taking arbitrary sums of Ps, as discussed earlier. We may use equation (9) in the most general case where $[\overline{P_\beta}]$ is a coarse graining of $[\overline{P_\alpha}]$.

A set of coarse-grained alternative histories is said to *decohere* when the off-diagonal elements of D are sufficiently small:

$$D([P_{\alpha'}], [P_\alpha]) \approx 0 \quad \text{for any } \alpha'_k \neq \alpha_k.$$

(10)

This is a generalization of the condition for the absence of interference in the two-slit experiment (approximate equality of the two sides of eq. 2). It is a sufficient (although not a necessary) condition for the validity of the purely diagonal formula

$$D([\overline{P_\beta}], [\overline{P_\beta}]) \approx \sum_{\substack{\text{all } P_\alpha \text{ not} \\ \text{fixed by } [\overline{P_\beta}]}} D([P_\alpha], [P_\alpha]).$$

(11)

The rule for when probabilities can be assigned to histories of the universe is then this: To the extent that a *set* of alternative histories decoheres, probabilities can be assigned to its individual members. The probabilities are the *diagonal* elements of D. Thus,

$$p([P_\alpha]) = D([P_\alpha], [P_\alpha])$$
$$= \text{Tr}\left[P_{\alpha_n}^n(t_n) \cdots P_{\alpha_1}^1(t_1) \rho P_{\alpha_1}^1(t_1) \cdots P_{\alpha_n}^n(t_n)\right] \quad (12)$$

when the set decoheres. We will frequently write $p(\alpha_n t_n, \ldots, \alpha_1 t_1)$ for these probabilities, suppressing the labels of the sets.

The probabilities defined by equation (12) obey the rules of probability theory as a consequence of decoherence. The principal requirement is that the probabilities be additive on "disjoint sets of the sample space." For histories this gives the sum rule

$$p([\overline{P_\beta}]) \approx \sum_{\substack{\text{all } P_\alpha \text{ not} \\ \text{fixed by } [\overline{P_\beta}]}} p([P_\alpha]). \quad (13)$$

These relate the probabilities for a set of histories to the probabilities for *all* coarser-grained sets that can be constructed from it. For example, the sum rule eliminating all projections at only one time is

$$\sum_{\alpha_k} p(\alpha_n t_n, \ldots, \alpha_{k+1} t_{k+1}, \alpha_k t_k, \alpha_{k-1} t_{k-1}, \ldots, \alpha_1 t_1)$$
$$\approx p(\alpha_n t_n, \ldots, \alpha_{k+1} t_{k+1}, \alpha_{k-1} t_{k-1}, \ldots, \alpha_1 t_1). \quad (14)$$

These rules follow trivially from equations (11) and (12). The other requirements from probability theory are that the probability of the whole sample space be unity, an easy consequence of equation (11) when complete coarse graining is performed, and that the probability for an empty set be zero, which means simply that the probability of any sequence containing a projection $P = 0$ must vanish, as it does.

The $p([P_\alpha])$ are *approximate* probabilities for histories, in the sense described in the section "Probability," up to the standard set by decoherence. Conversely, if a given standard for the probabilities is required by their use, it can be met by coarse graining until equations (10) and (13) are satisfied at the requisite level.

Further coarse graining of a decoherent set of alternative histories produces another set of decoherent histories, since the probability sum rules continue to be obeyed. That is illustrated in figure 2, which makes it clear that in a progression from the trivial completely coarse graining to a completely fine graining, there are sets of histories where further fine graining always results in loss of decoherence. These are the *maximal* sets of alternative decohering histories.

These rules for probability exhibit another important feature: the operators in equation (12) are time ordered. Were they not time ordered (zig-zags) we could have assigned non-zero probabilities to conflicting alternatives at the same time. The time ordering thus expresses causality in quantum mechanics, a notion that is appropriate here because of the approximation of fixed background spacetime. The time ordering is related as well to the "arrow of time" in quantum mechanics, which we discuss below.

Given this discussion, the *fundamental formula* of quantum mechanics may be reasonably taken to be

$$D([P_{\alpha'}], [P_\alpha]) \approx \delta_{\alpha'_1 \alpha_1} \cdots \delta_{\alpha'_n \alpha_n} p([P_\alpha]) \qquad (15)$$

for all $[P_\alpha]$ in a set of alternative histories. Vanishing of the off-diagonal elements of D gives the rule for when probabilities may be consistently assigned. The diagonal elements give their values.

We could have used a weaker condition than equation (10) as the definition of decoherence, namely the necessary condition for the validity of the sum rules of probability theory given in equation (11):

$$D([P_\alpha], [P_{\alpha'}]) + D([P_{\alpha'}], [P_\alpha]) \approx 0 \qquad (16)$$

for any $\alpha'_k \neq \alpha_k$, or equivalently

$$\mathrm{Re}\{D([P_\alpha], [P_{\alpha'}])\} \approx 0. \qquad (17)$$

This is the condition used by Griffiths (1984) as the requirement for "consistent histories." However, while, as we shall see, it is easy to identify physical situations in which the off-diagonal elements of D approximately vanish as the result of coarse graining, it is hard to think of a general mechanism that suppresses only their real parts. In the usual analysis of measurement, the off-diagonal parts of D approximately vanish. We shall, therefore, explore the stronger condition of equation (10) in what follows. That difference should not obscure the fact that in this part of our work we have reproduced what is essentially the approach of Griffiths (1984), extended by Omnès (1988a, 1988b, 1988c).

(D) PREDICTION AND RETRODICTION

Decoherent sets of histories are what we may discuss in quantum mechanics, for they may be assigned probabilities. Decoherence thus generalizes and replaces the notion of "measurement," which served this role in the Copenhagen interpretations. Decoherence is a more precise, more objective, more observer-independent idea. For example, if their associated histories decohere, we may assign probabilities to various values of reasonable scale density fluctuations in the early universe whether or not anything like a "measurement" was carried out on them and certainly whether or not there was an "observer" to do it. We shall return to a specific discussion of typical measurement situations in a later section.

The joint probabilities $p(\alpha_n t_n, \ldots, \alpha_1 t_1)$ for the individual histories in a decohering set are the raw material for prediction and retrodiction in quantum cosmology. From them, the relevant conditional probabilities may be computed. The conditional probability of one subset $\{\alpha_i t_i\}$, given the rest $\overline{\{\alpha_i t_i\}}$, is

$$p(\{\alpha_i t_i\} \mid \overline{\{\alpha_i t_i\}}) = \frac{p(\alpha_n t_n, \ldots, \alpha_1 t_1)}{p(\overline{\{\alpha_i t_i\}})}. \qquad (18)$$

For example, the probability for *predicting* alternatives α_{k+1}, $\ldots, \alpha_n$, given that the alternatives $\alpha_1, \ldots, \alpha_k$ have already happened, is

$$p(\alpha_n t_n, \ldots, \alpha_{k+1} t_{k+1} \mid \alpha_k t_k, \ldots, \alpha_1 t_1) = \frac{p(\alpha_n t_n, \ldots, \alpha_1 t_1)}{p(\alpha_k t_k, \ldots, \alpha_1 t_1)}. \qquad (19)$$

The probability that $\alpha_{n-1}, \ldots, \alpha_1$ happened in the *past*, given present data summarized by an alternative α_n at the present time t_n, is

$$p(\alpha_{n-1} t_{n-1}, \ldots, \alpha_1 t_1 \mid \alpha_n t_n) = \frac{p(\alpha_n t_n, \ldots, \alpha_1 t_1)}{p(\alpha_n t_n)}. \qquad (20)$$

Decoherence ensures that the probabilities defined by equations (18)–(20) will approximately add to unity when summed over all remaining alternatives, because of equation (14).

Despite the similarity between equations (19) and (20), there are differences between prediction and retrodiction. Future predictions can all be obtained from an effective density matrix summarizing information about what has happened. If ρ_{eff} is defined by

$$\rho_{\text{eff}} = \frac{P_{\alpha_k}^k(t_k) \cdots P_{\alpha_1}^1(t_1) \rho P_{\alpha_1}^1(t_1) \cdots P_{\alpha_k}^k(t_k)}{\text{Tr}[P_{\alpha_k}^k(t_k) \cdots P_{\alpha_1}^1(t_1) \rho P_{\alpha_1}^1(t_1) \cdots P_{\alpha_k}^k(t_k)]}, \qquad (21)$$

then

$$p(\alpha_n t_n, \ldots, \alpha_{k+1} t_{k+1} \mid \alpha_k t_k, \ldots, \alpha_1 t_1)$$
$$= \mathrm{Tr}[P^n_{\alpha_n}(t_n) \ldots P^{k+1}_{\alpha_{k+1}}(t_{k+1}) \rho_{\mathrm{eff}} P^{k+1}_{\alpha_{k+1}}(t_{k+1}) \ldots P^n_{\alpha_n}(t_n)].$$
$$(22)$$

By contrast, there is no effective density matrix representing present information from which probabilities for the past can be derived. As equation (20) shows, history requires knowledge of both present data and the initial condition of the universe. ~643~

Prediction and retrodiction differ in another way. Because of the cyclic property of the trace in equation (8), *any* final alternative decoheres and a probability can be predicted for it. By contrast, we expect only certain variables to decohere in the past, appropriate to present data and the initial ρ. As the alternative histories of the electron in the two-slit experiment illustrate, there are many kinds of alternatives in the past for which the assignment of probabilities is prohibited in quantum mechanics. For those sets of alternatives that do decohere, the decoherence and the assigned probabilities typically will be approximate in the sense described in the "Probability" section. It is unlikely, for example, that the initial state of the universe is such that the interference is exactly zero between two past positions of the sun in the sky.

These differences between prediction and retrodiction are aspects of the arrow of time in quantum mechanics. Mathematically they are consequences of the time ordering in equation (8) or (12). This time ordering does not mean that quantum mechanics singles out an absolute direction in time. Field theory is invariant under CPT. Performing a CPT transformation on equation (8) or (12) results in an equivalent expression in which the CPT-transformed ρ is assigned to the far future and the CPT-transformed projections are anti-time-ordered. Either time ordering can, therefore, be used;[10] the important point is that there is a knowable Heisenberg ρ from which probabilities can be predicted. It is by convention that we think of it as an "initial condition," with the projections in increasing time order from the inside out in equation (8) or (12).

While the formalism of quantum mechanics allows the universe to be discussed with either time ordering, the physics of the universe is time asymmetric, with a simple condition in what we call "the past." For example, the indicated present homogeneity of the thermodynamic arrow of time can be traced to the near homogeneity of the "early" universe implied by ρ and the implication that the progenitors of approximately isolated subsystems started out far from equilibrium at "early" times.

Much has been made of the updating of the fundamental probability formula in equation (19) and in equations (21) and (22). By utilizing equation (21) the process of prediction may be organized so that for each time there is a ρ_{eff} from which probabilities for the future may be calculated. The action of each projection, P, on both sides of ρ in equation (21) along with the division by the appropriate normalizing factor is then sometimes called the "reduction of the wave packet." But this updating of probabilities is no different from the classical reassessment of probabilities that occurs after new information is obtained. In a sequence of horse races, the joint probability for the winners

---

[10]It has been suggested—see Hartle (1988a, 1988b, 1989b, 1990a)—that for application to highly quantum-mechanical spacetime, as in the very early universe, quantum mechanics should be generalized to yield a framework in which both time orderings are treated simultaneously in the sum-over-histories approach. This involves including both $\exp(iS)$ and $\exp(-iS)$ for each history and has as a consequence an evolution equation (the Wheeler–DeWitt equation) that is of second order in the time variable. The suggestion is that the two time orderings decohere when the universe is large and spacetime classical, so that the usual framework with just one ordering is recovered.

of eight races is converted, after the winners of the first three are known, into a reassessed probability for the remaining five races by exactly this process. The main thing is that, because of decoherence, the sum rules for probabilities are obeyed; once that is true, reassessment of probabilities is trivial.

The only non-trivial aspect of the situation is the choice of the string of Ps in equation (8) giving a decoherent set of histories.

(E) BRANCHES (ILLUSTRATED BY A PURE ρ)

Decohering sets of alternative histories give a definite meaning to Everett's "branches." For a given such set of histories, the exhaustive set of $P_{\alpha_k}^k$ at each time t_k corresponds to a branching.

To illustrate this even more explicitly, consider an initial density matrix that is a pure state, as in typical proposals for the wave function of the universe:

$$\rho = |\Psi\rangle\langle\Psi|. \tag{23}$$

The initial state may be decomposed according to the projection operators that define the set of alternative histories:

$$
\begin{aligned}
|\Psi\rangle &= \sum_{\alpha_1,...,\alpha_n} P_{\alpha_n}^n(t_n) \cdots P_{\alpha_1}^1(t_1)|\Psi\rangle \\
&\equiv \sum_{\alpha_1,...,\alpha_n} |[P_\alpha], \Psi\rangle.
\end{aligned}
\tag{24}
$$

The states $|[P_\alpha], \Psi\rangle$ are approximately orthogonal as a consequence of their decoherence,

$$\langle[P_{\alpha'}], \Psi|[P_\alpha], \Psi\rangle \approx 0 \quad \text{for any } \alpha_k' \neq \alpha_k. \tag{25}$$

Equation (25) is just a re-expression of equation (10), given equation (23).

When the initial density matrix is pure, it is easily seen that some coarse graining in the present is always needed to achieve decoherence in the past. If the $P_{\alpha_n}^n(t_n)$ for the last time t_n in equation (8) were all projections onto pure states, D would factor for a pure ρ and could never satisfy equation (10), except for certain special kinds of histories described near the end of the section "Maximal Sets of Decohering Histories," in which decoherence is automatic, independent of ρ. Similarly, it is not difficult to show that some coarse graining is required at any time in order to have decoherence of previous alternatives, with the same set of exceptions.

After normalization, the states $|[P_\alpha], \Psi\rangle$ represent the individual histories or individual branches in the decohering set. We may, as for the effective density matrix of subsection (D), summarize present information for prediction just by giving one of these states, with projections up to the present.

(F) SETS OF HISTORIES WITH THE SAME PROBABILITIES

If the projections P are not restricted to a particular class (such as projections onto ranges of Q^i variables), so that coarse-grained histories consist of arbitrary exhaustive families of projection operators, then the problem of exhibiting the decohering sets of strings of projections arising from a given ρ is a purely algebraic one. Assume, for example, that the initial condition is known to be a pure state as in equation (23). The problem of finding ordered strings of exhaustive sets of projections $[P_\alpha]$ so that the histories $P_{\alpha_n}^n \cdots P_{\alpha_1}^1 |\Psi\rangle$ decohere according to equation (25) is purely algebraic and involves just subspaces of Hilbert space. The problem is the same for one vector $|\Psi\rangle$ as for any other. Indeed, using subspaces that are *exactly* orthogonal, we may identify sequences that *exactly* decohere.

However, it is clear that the solution of the mathematical problem of enumerating the sets of decohering histories of a given Hilbert space has no physical content by itself. No description of the histories has been given. No reference has been made to a theory of the fundamental interactions. No distinction has been made between one vector in Hilbert space as a theory of the initial condition and any other. The resulting probabilities, which can be calculated, are merely abstract numbers.

We obtain a description of the sets of alternative histories of the universe when the operators corresponding to the fundamental fields are identified. We make contact with the theory of the fundamental interactions if the evolution of these fields is given by a fundamental Hamiltonian. Different initial vectors in Hilbert space will then give rise to decohering sets having different descriptions in terms of the fundamental fields. The probabilities acquire physical meaning.

Two different simple operations allow us to construct from one set of histories another set with a *different description* but the *same probabilities*. First consider unitary transformations of the Ps that are constant in time and leave the initial ρ fixed:

$$\rho = U \rho U^{-1}, \qquad (26)$$

$$\tilde{P}_\alpha^k(t) = U P_\alpha^k(t) U^{-1}. \qquad (27)$$

If ρ is pure, there will be very many such transformations; the Hilbert space is large and only a single vector is fixed. The sets of histories made up from the $\{\tilde{P}_\alpha^k\}$ will have an identical decoherence functional to the sets constructed from the corresponding $\{P_\alpha^k\}$. If one set decoheres, the other will, and the probabilities for the individual histories will be the same.

In a similar way, decoherence and probabilities are invariant under arbitrary reassignments of the times in a string of Ps (as long as they continue to be ordered), with the projection operators at the altered times unchanged as operators in Hilbert space. This is because in the Heisenberg picture every projection is at any time a projection operator for *some* quantity.

The histories arising from constant unitary transformations or from reassignment of times of a given set of Ps will, in general, have very different descriptions in terms of fundamental fields from that of the original set. We are considering transformations such as equation (27) in an active (or alibi) sense so that the field operators and Hamiltonian are unchanged. (The passive [or alias] transformations, in which these are transformed, are easily understood.) A set of projections onto the ranges of field values in a spatial region is generally transformed by equation (27) or by any reassignment of the times into an extraordinarily complicated combination of all fields and all momenta at all positions in the universe! Histories consisting of projections onto values of similar quantities at different times can thus become histories of very different quantities at various other times.

In ordinary presentations of quantum mechanics, two histories with different descriptions can correspond to physically distinct situations because it is presumed that various different Hermitian combinations of field operators are potentially measurable by different kinds of external apparatus. In quantum cosmology, however, apparatus and system are considered together and the notion of physically distinct situations may have a different character.

V. The Origins of Decoherence

What are the features of coarse-grained sets of histories that decohere, given the ρ and H of the universe? In seeking to answer this question it is important to keep in mind the basic aspects of the theoretical framework on which decoherence depends. Decoherence of a set of alternative histories is not a property of their operators *alone*. It depends on the relations of those operators to the density matrix ρ. Given ρ, we could, in principle, *compute* which sets of alternative histories decohere.

We are not likely to carry out a computation of all decohering sets of alternative histories for the universe, described in terms of the fundamental fields, anytime in the near future, if ever. However, if we focus attention on coarse grainings of particular variables, we can exhibit widely occurring mechanisms by which they decohere in the presence of the actual ρ of the universe. We have mentioned in the subsection "(C) Decohering Histories" that decoherence is automatic if the projection operators P refer only to one time; the same would be true even for different times if all the Ps commuted with one another. Of course, in cases of interest, each P typically factors into commuting projection operators, and the factors of Ps for different times often fail to commute with one another—for example, factors that are projections onto related ranges of values of the same Heisenberg operator at different times. However, these non-commuting factors may be correlated, given ρ, with other projection factors that do commute or, at least, effectively commute inside the trace with the density matrix ρ in equation (8) for the decoherence functional. In fact, these other projection factors may commute with all the subsequent Ps and thus allow themselves to be moved to the outside of the trace formula. When all the non-commuting factors are correlated in this manner with effectively commuting ones, then the off-diagonal terms in the decoherence functional vanish—in other words, decoherence results. Of course, all this behavior may be approximate, resulting in approximate decoherence.

This type of situation is fundamental in the interpretation of quantum mechanics. Non-commuting quantities, say at different times, may be correlated with commuting or effectively commuting quantities because of the character of ρ and H, and thus produce decoherence of strings of Ps despite their non-commutation. For a pure ρ, for example, the behavior of the effectively commuting variables leads to the orthogonality of the branches of the state $|\Psi\rangle$, as defined in equation (24). We shall see that correlations of this character are central to understanding historical records (see "Branch Dependence") and measurement situations (see the section "Measurement Situations").

As an example of decoherence produced by this mechanism, consider a coarse-grained set of histories defined by time sequences of alternative approximate localizations of a massive body such as a planet or even a typical interstellar dust grain. As shown by Joos and Zeh (1985), even if the successive localizations are spaced as closely as a nanosecond, such histories decohere as a consequence of scattering by the 3° cosmic background radiation (if for no other reason). Different positions become correlated with nearly orthogonal states of the photons. More importantly, each alternative sequence of positions becomes correlated with a different orthogonal state of the photons at the final time. This accomplishes the decoherence and we may loosely say that such histories of the position of a massive body are "decohered" by interaction with the photons of the background radiation.

Other specific models of decoherence have been discussed by many authors, among them Joos and Zeh (1985), Caldeira and Leggett (1983), and Zurek (1984). Typically these discussions have focused on a coarse graining that involves only certain variables analogous to the position variables above. Thus the emphasis is on particular non-commuting factors of the projection operators and not on correlated operators that may be accomplishing the approximate decoherence. Such coarse grainings do not, in general, yield the most refined approximately decohering sets of histories, since one could include projections onto ranges of values of the correlated operators without losing the decoherence.

The simplest model consists of a single oscillator interacting bilinearly with a large number of others, and a coarse graining which involves only the coordinates of the special oscillator. Let x be the coordinate of the special oscillator, M its mass, ω_R its frequency renormalized by its interactions with the others, and S_{free} its free action. Consider the special case where the density matrix of the whole system, referred to an initial time, factors into the product of a density matrix $\overline{\rho}(x', x)$ of the distinguished oscillator and another for the rest. Then, generalizing slightly a treatment of Feynman and Vernon (1963), we can write D defined by equation (7) as

$$D([\Delta_{\alpha'}], [\Delta_\alpha]) = \int_{[\Delta_{\alpha'}]} \delta x'(t) \int_{[\Delta_\alpha]} \delta x(t) \delta(x'_f - x_f)$$
$$\exp\left\{ i\big(S_{\text{free}}[x'(t)] - S_{\text{free}}[x(t)] + W[x'(t), x(t)]\big)/\hbar \right\} \overline{\rho}(x'_0, x_0),$$
$$(28)$$

the intervals $[\Delta_\alpha]$ referring here only to the variables of the distinguished oscillator. The sum over the rest of the oscillators has been carried out and is summarized by the Feynman–Vernon influence functional $\exp(iW[x'(t), x(t)])$. The remaining sum over $x'(t)$ and $x(t)$ is as in equation (7).

The case where the other oscillators are in an initial thermal distribution has been extensively investigated by Caldeira and Leggett (1983). In the simple limit of a uniform continuum of oscillators cut off at frequency Ω and in the Fokker–Planck limit of $kT \gg \hbar\Omega \gg \hbar\omega_R$, they find

$$W[x'(t), x(t)] = -M\gamma \int dt \left[x'\dot{x}' - x\dot{x} + x'\dot{x} - x\dot{x}'\right]$$
$$+ i\frac{2M\gamma kT}{\hbar} \int dt \left[x'(t) - x(t)\right]^2,$$
$$(29)$$

where γ summarizes the interaction strengths of the distinguished oscillator with its environment. The real part of W contributes dissipation to the equations of motion. The imaginary part squeezes the trajectories $x(t)$ and $x'(t)$ together, thereby providing approximate decoherence. Very roughly, primed and unprimed position intervals differing by distances d on opposite sides of the trace in equation (28) will decohere when spaced in time by intervals

$$t \gtrsim \frac{1}{\gamma} \left[\left(\frac{\hbar}{\sqrt{2MkT}} \right) \cdot \left(\frac{1}{d} \right) \right]^2 . \qquad (30)$$

As stressed by Zurek (1984), for typical macroscopic parameters this minimum time for decoherence can be many orders of magnitude smaller than a characteristic dynamical time, say the damping time $1/\gamma$. (The ratio is around 10^{-40} for $M \sim$ gm, $T \sim 300$ K, and $d \sim$ cm!)

The behavior of a coarse-grained set of alternative histories based on projections, at times spaced far enough apart for decoherence, onto ranges of values of x alone, is then roughly classical in that the successive ranges of positions follow roughly classical orbits, but with the pattern of classical correlation disturbed by various effects, especially (a) the effect of quantum spreading of the x-coordinate, (b) the effect of quantum fluctuations of the other oscillators, and (c) classical statistical fluctuations, which are lumped with (b) when we use the fundamental formula. We see that the larger the mass M, the shorter the decoherence time and the more the x-coordinate resists the various challenges to its classical behavior.

What the above models convincingly show is that decoherence will be widespread in the universe for certain familiar "classical" variables. The answer to Fermi's question to one of us of why we don't see Mars spread out in a quantum superposition of different positions in its orbit is that such a superposition would rapidly decohere. We now proceed to a more detailed discussion of such decoherence.

VI. Quasiclassical Domains

As observers of the universe, we deal with coarse grainings that are appropriate to our limited sensory perceptions, extended by instruments, communication, and records, but in the end characterized by a great amount of ignorance. Yet we have the impression that the universe exhibits a finer-grained set of decohering histories, independent of us, defining a sort of "classical domain," governed largely by classical laws, to which our senses are adapted while dealing with only a small part of it. No such coarse graining is determined by pure quantum theory alone. Rather, like decoherence, the existence of a quasiclassical domain in the universe must be a consequence of its initial condition and the Hamiltonian describing its evolution.

~ 653 ~

Roughly speaking, a quasiclassical domain should be a set of alternative decohering histories, maximally refined consistent with decoherence, with its individual histories exhibiting as much as possible patterns of classical correlation in time. Such histories cannot be *exactly* correlated in time according to classical laws because sometimes their classical evolution is disturbed by quantum events. There are no classical domains, only quasiclassical ones.

We wish to make the question of the existence of one or more quasiclassical domains into a *calculable* question in quantum cosmology, and for this we need criteria to measure how close a set of histories comes to constituting a "classical domain." We have not solved this problem to our satisfaction, but, in the next few sections, we discuss some ideas that may contribute toward its solution.

VII. Maximal Sets of Decohering Histories

Decoherence results from coarse graining. As described in the subsection "(B) Histories" and figure 2, coarse grainings can be put into a partial ordering with one another. A set of alternative histories is a coarse graining of a finer set if all the exhaustive sets of projections $\{P_\alpha^k\}$ making up the coarser set of histories are obtained by partial sums over the projections making up the finer set of histories.

Maximal sets of alternative decohering histories are those for which there are no finer-grained sets that are decoherent. It is desirable to work with maximal sets of decohering alternative histories because they are not limited by the sensory capacity of any set of observers—they can cover phenomena in all parts of the universe and at all epochs that could be observed, whether or not any observer was present. Maximal sets are the most refined descriptions of the universe that may be assigned probabilities in quantum mechanics.

The class of maximal sets possible for the universe depends, of course, on the completely fine-grained histories that are presented by the actual quantum theory of the universe. If we utilize to the full, at each moment of time, all the projections permitted by transformation theory, which gives quantum mechanics its protean character, then there is an infinite variety of completely fine-grained sets, as illustrated in figure 2. However, were there some fundamental reason to restrict the completely fine-grained sets, as would be the case if sum-over-histories quantum mechanics were fundamental, then the class of maximal sets would be smaller, as illustrated in figure 4. We shall proceed as if all fine grainings are allowed.

If a full correlation exists between a projection in a coarse graining and another projection not included, then the finer graining including both still defines a decoherent set of histories. In a maximal set of decoherent histories, both correlated projections must be included if either one is included. Thus, in the mechanism of decoherence discussed in the section "The Origins

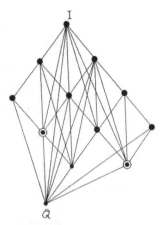

Figure 4. If the completely fine-grained histories arise from a single complete set of observables, say the set Q of field variables Q^i at each point in space and every time, then the possible coarse-grained histories will be a subset of those illustrated in figure 2. Maximal sets can still be defined but will, in general, differ from those of figure 2.

of Decoherence," projections onto the correlated orthogonal states of the 3 K photons are included in the maximal set of decohering histories along with the positions of the massive bodies. Any projections defining historical records, such as we shall describe in the section "Branch Dependence," or values of measured quantities, such as we shall describe in the section "Measurement Situations," must similarly be included in a maximal set.

More information about the initial ρ and H is contained in the probabilities of a finer-grained set of histories than in those of a coarser-grained set. It would be desirable to have a quantitative measure of *how much* more information is obtained in a further fine graining of a coarse-grained set of alternative histories. Such a quantity would then measure how much closer a decoherent fine graining comes to maximality in a physically relevant sense.

We shall discuss a quantity that, while not really a measure of maximality, is useful in exploring some aspects of it. In order to construct that quantity, the usual entropy formula is applied to sets of alternative decohering *histories* of the universe, rather than, as more usually, alternatives at a single time. We make use of the coarse-grained density matrix $\tilde{\rho}$ defined using the methods of Jaynes,[11] *but generalized to take account of the density matrix of the universe and applied to the probabilities for histories.* The density matrix $\tilde{\rho}$ is constructed by maximizing the entropy functional

$$S(\tilde{\rho}) = -\text{Tr}(\tilde{\rho}\log\tilde{\rho}) \tag{31}$$

over all density matrices $\tilde{\rho}$ that satisfy the constraints ensuring that each

$$\text{Tr}\left[P_{\alpha_n'}^n(t_n)\cdots P_{\alpha_1'}^1(t_1)\tilde{\rho}P_{\alpha_1}^1(t_1)\cdots P_{\alpha_n}^n(t_n)\right] \tag{32}$$

has the same value it would have had when computed with the density matrix of the universe, ρ, for a given set of coarse-grained histories. The density matrix $\tilde{\rho}$ thus reproduces the decoherence functional for this set of histories, and in particular their probabilities, but possesses as little information as possible beyond those properties.

A fine graining of a set of alternative histories leads to *more* conditions on $\tilde{\rho}$ of the form of equation (32) than in the coarser-grained set. In non-trivial cases $S(\tilde{\rho})$ is, therefore, lowered and $\tilde{\rho}$ becomes closer to ρ.

If the insertion of apparently new Ps into a chain is redundant, then $S(\tilde{\rho})$ will not be lowered. A simple example will help to illustrate this. Consider the set of histories consisting of projections $P_{\alpha_m}^m(t_m)$ which project onto an orthonormal basis for Hilbert space at one time, t_m. Trivial further decoherent fine grainings can be constructed as follows: at each other time t_k introduce a set of projections $P_{\alpha_k}^k(t_k)$ that, through the equations of motion, are identical operators in Hilbert space

---

[11]See, e.g., the papers reprinted in Rosenkrantz (1983) and Hobson (1971).

to the set $\{P_{\alpha_m}^m(t_m)\}$. In this way, even though we are going through the motions of introducing a completely fine-grained set of histories covering all the times, we are really just repeating the projections $P_{\alpha_m}^m(t_m)$ over and over again. We thus have a completely fine-grained set of histories that, in fact, consists of just one fine-grained set of projections and decoheres exactly because there is only one such set. Indeed, in terms of $S(\tilde{\rho})$ it is no closer to maximality than the set consisting of $P_{\alpha_m}^m(t_m)$ at one time. The quantity $S(\tilde{\rho})$ thus serves to identify such trivial refinements, which amount to redundancy in the conditions given in equation (32).

We can generalize the example in an interesting way by constructing the special kinds of histories mentioned after equation (25). We take t_m to be the final time and then adjoin, at earlier and earlier times, a succession of progressive coarse grainings of the set $\{P_{\alpha_m}^m(t_m)\}$. Thus, as time moves forward, the only projections are finer and finer grainings terminating in the one-dimensional $P_{\alpha_m}^m(t_m)$. We thus have again a set of histories in which decoherence is automatic and independent of the character of ρ, and for which $S(\tilde{\rho})$ has the same value it would have had if only conditions at the final time had been considered.

In a certain sense, $S(\tilde{\rho})$ for histories can be regarded as decreasing with time. If we consider $S(\tilde{\rho})$ for a string of alternative projections up to a certain time t_n, as in equation (32), and then adjoin an additional set of projections for a later time, the number of conditions on ρ is increased and thus the value of $S(\tilde{\rho})$ is decreased (or, in trivial cases, unchanged). That is natural, since $S(\tilde{\rho})$ is connected with the lack of information contained in a set of histories and that information increases with non-trivial fine graining of the histories, no matter what the times for which the new Ps are introduced. (In some related problems, a quantity like S that keeps decreasing as a result of adjoining projections at later times can be converted into an increasing quantity by adding an algorithmic complexity term; see Zurek [1989].)

The quantity $S(\tilde{\rho})$ is closely related to other fundamental quantities in physics. One can show, for example, that when *used with the* ρ_{eff} *representing present data and with alternatives at a single time*, these techniques give a unified and generalized treatment of the variety of coarse grainings commonly introduced in statistical mechanics; and, as Jaynes and others have pointed out, the resulting $S(\tilde{\rho})$s are the physical entropies of statistical mechanics. Here, however, these techniques are applied to time histories and the initial condition is utilized. The quantity $S(\tilde{\rho})$ is also related to the notion of thermodynamic depth currently being investigated by Lloyd.

VIII. Classicity

Some maximal sets will be more nearly classical than others. The more nearly classical sets of histories will contain projections (onto related ranges of values) of operators, for different times, that are connected to one another by the unitary transformations $e^{-iH(\Delta t)}$ and that are correlated for the most part along classical paths, with probabilities near zero and one for the successive projections. This pattern of classical correlation may be disturbed by the inclusion, in the maximal set of projection operators, of other variables which do not behave in this way (as in measurement situations to be described later). The pattern may also be disturbed by quantum spreading and by quantum and classical fluctuations, as described in connection with the oscillator example treated in the section "The Origins of Decoherence." Thus we can, at best, deal with *quasiclassical* maximal sets of alternative decohering histories, with trajectories that split and fan out as a result of the processes that make the decoherence possible. As we stressed earlier, there are no classical domains, only quasiclassical ones.

The impression that there is something like a classical domain suggests that we try to define quasiclassical domains precisely by searching for a measure of classicity for each of the maximal sets of alternative decohering histories and concentrating on the one (or ones) with maximal classicity. Such a measure would be applied to the elements of D and the corresponding coarse graining. It should favor predictability, involving patterns of classical correlation as described above. It should also favor maximal sets of alternative decohering histories that are relatively fine-grained, as opposed to those which had to be carried to very coarse graining before they would give decoherence. We are searching for such a measure. It should provide a precise and quantitative meaning to the notion of quasiclassical domain.

~ 659 ~

IX. Quasiclassical Operators

What are the projection operators that specify the coarse graining of a maximal set of alternative histories with high classicity, which defines a quasiclassical domain? They will include, as mentioned above, projections onto comparable ranges of values of certain operators at sequences of times, obeying roughly classical equations of motion, subject to fluctuations that cause their trajectories to fan out from time to time. We can refer to these operators, which habitually decohere, as "quasiclassical operators." What these quasiclassical operators are, and how many of them there are, depends not only on H and ρ but also on the epoch, on the spatial region, and on previous branchings.

We can understand the origin of at least some quasiclassical operators in reasonably general terms as follows: In the earliest instants of the universe the operators defining spacetime on scales well above the Planck scale emerge from the quantum fog as quasiclassical.[12] Any theory of the initial condition that does not imply this is simply inconsistent with observation in a manifest way. The background spacetime thus defined obeys the Einstein equation. Then, where there are suitable conditions of low temperature, etc., various sorts of hydrodynamic variables may emerge as quasiclassical operators. These are integrals over suitable small volumes of densities of conserved or nearly conserved quantities. Examples are densities of energy, momentum, baryon number, and, in later epochs, nuclei, and even chemical species. The sizes of the volumes are limited above by maximality and are limited below by classicity because they require sufficient "inertia" to enable them to resist deviations from predictability caused by their interactions with one another, by quantum spreading and by the quantum and statistical fluctuations resulting from interactions with the rest of the universe. Suitable integrals of densities of approximately conserved quantities are thus candidates for habitually decohering quasiclassical operators. Field theory is local, and it is an interesting question whether that locality somehow picks out local densities as the source of habitually decohering quantities. It is hardly necessary to note that such hydrodynamic variables are among the principal variables of classical physics.[13]

In the case of densities of conserved quantities, the integrals would not change at all if the volumes were infinite. For smaller volumes we expect approximate persistence. When, as in hydrodynamics, the rates of change of the integrals form part of an approximately closed system of equations of motion, the resulting evolution is just as classical as in the case of persistence.

---

[12]See, e.g., Joos (1986), Zeh (1986), Kiefer (1987), Halliwell (1989), and Fukuyama and Morikawa (1989).

[13]For discussion of how such hydrodynamic variables are distinguished in non-equilibrium statistical mechanics in not unrelated ways see, e.g., Kadanoff and Martin (1963), Forster (1975), and Lebowitz (1986).

X. Branch Dependence

As the discussion in the sections "The Origins of Decoherence" and "Quasiclassical Operators" shows, physically interesting mechanisms for decoherence will operate differently in different alternative histories for the universe. For example, hydrodynamic variables defined by a relatively small set of volumes may decohere at certain locations in spacetime in those branches where a gravitationally condensed body (e.g., the earth) actually exists, and may not decohere in other branches where no such condensed body exists at that location. In the latter branch there simply may be not enough "inertia" for densities defined with too small volumes to resist deviations from predictability. Similarly, alternative spin directions associated with Stern–Gerlach beams may decohere for those branches on which a photographic plate detects their beams and not in a branch where they recombine coherently instead. There are no variables that are expected to decohere universally. Even the mechanisms causing spacetime geometry at a given location to decohere on scales far above the Planck length cannot necessarily be expected to operate in the same way on a branch where the location is the center of a black hole as on those branches where there is no black hole nearby.

How is such "branch dependence" described in the formalism we have elaborated? It is not described by considering histories where the *set* of alternatives at one time (the k in a set of P_α^k) depends on *specific* alternatives (the αs) of sets of earlier times. Such dependence would destroy the derivation of the probability sum rules from the fundamental formula. However, there is no such obstacle to the set of alternatives at one time depending on the *sets* of alternatives at all previous times. It is by exploiting this possibility, together with the possibility of present records of past events, that we can correctly describe the sense in which there is branch dependence of decoherence, as we shall now discuss.

A record is a present alternative that is, with high probability, correlated with an alternative in the past. The construction of the relevant probabilities was discussed in the section "Decoherent Sets of Histories," including their dependence on the initial condition of the universe (or at least on information that effectively bears on that initial condition). The subject of history is most honestly described as the construction of probabilities for the past, given such records. Even non-commuting alternatives such as a position and its momentum at different, even nearby, times may be stored in presently commuting record variables.

The branch dependence of histories becomes explicit when sets of alternatives are considered that include records of specific events in the past. To illustrate this, consider the example above, where different sorts of hydrodynamic variables might decohere or not depending on whether there was a gravitational condensation. The set of alternatives that decohere must refer both to the records of the condensation *and* to hydrodynamic variables. Hydrodynamic variables with smaller volumes would be part of the subset with the record that the condensation took place and vice versa.

The branch dependence of decoherence provides the most direct argument against the position that a classical domain should simply be *defined* in terms of a certain set of variables (e.g., values of spacetime averages of the fields in the classical action). There are unlikely to be any physically interesting variables that decohere independent of circumstance.

XI. Measurement Situations

When a correlation exists between the ranges of values of two operators of a quasiclassical domain, there is a *measurement situation*. From a knowledge of the value of one, the value of the other can be deduced because they are correlated with probability near unity. Any such correlation exists in some branches of the universe and not in others; for example, measurements in a laboratory exist only in those branches where the laboratory was actually constructed!

We use the term "measurement situation" rather than "measurement" for such correlations to stress that nothing as sophisticated as an "observer" need be present for them to exist. If there are many significantly different quasiclassical domains, different measurement situations may be exhibited by each one.

When the correlation we are discussing is between the ranges of values of two quasiclassical operators that *habitually* decohere, as discussed in the section "Quasiclassical Operators," we have a measurement situation of a familiar classical kind. However, besides the quasiclassical operators, the highly classical maximal sets of alternative histories of a quasiclassical domain may include *other* operators having ranges of values strongly correlated with the quasiclassical ones at particular times. Such operators, not normally decohering, are, in fact, included among the decohering set only by virtue of their correlation with a habitually decohering one. In this case we have a measurement situation of the kind usually discussed in quantum mechanics. Suppose, for example, in the inevitable Stern–Gerlach experiment, that σ_z of a spin-$\frac{1}{2}$ particle is correlated with the orbit of an atom in an inhomogeneous magnetic held. If the two orbits decohere because of interaction with something else (the atomic excitations in a photographic plate, for example), then the spin direction will be included in the maximal set of decoherent histories, fully correlated with the decohering orbital directions. The spin direction is thus measured.

The recovery of the Copenhagen rule for when probabilities may be assigned is immediate. Measured quantities are correlated with decohering histories. Decohering histories can be assigned

- 663 -

probabilities. Thus in the two-slit experiment (fig. 1), when the electron interacts with an apparatus that determines which slit it passed through, it is the decoherence of the alternative configurations of the apparatus that enables probabilities to be assigned for the electron.

Correlation between the ranges of values of operators of a quasiclassical domain is the *only* defining property of a measurement situation. Conventionally, measurements have been characterized in other ways. Essential features have been seen to be irreversibility, amplification beyond a certain level of signal-to-noise, association with a macroscopic variable, the possibility of further association with a long chain of such variables, and the formation of enduring records. Efforts have been made to attach some degree of precision to words such as "irreversible," "macroscopic," and "record," and to discuss what level of "amplification" needs to be achieved.[14] While such characterizations of measurement are difficult to define precisely,[15] some can be seen in a rough way to be consequences of the definition that we are attempting to introduce here, as follows.

Correlation of a variable with the quasiclassical domain (actually, inclusion in its set of histories) accomplishes the amplification beyond noise and the association with a macroscopic variable that can be extended to an indefinitely long chain of such variables. The

---

[14]For an interesting effort at precision see Daneri, Loinger, and Prosperi (1962).

[15]An example of this occurs in the case of "null measurements" discussed by Renninger (1960), Dicke (1981), and others. An atom decays at the center of a spherical cavity. A detector which covers all but a small opening in the sphere does not register. We conclude that we have measured the direction of the decay photon to an accuracy set by the solid angle subtended by the opening. Certainly there is an interaction of the electromagnetic field with the detector, but did the escaping photon suffer an "irreversible act of amplification"? The point in the present approach is that the *set* of alternatives, detected and not detected, exhibits decoherence because of the place of the detector in the universe.

relative predictability of the classical world is a generalized form of record. The approximate constancy of, say, a mark in a notebook is just a special case; persistence in a classical orbit is just as good.

Irreversibility is more subtle. One measure of it is the cost (in energy, money, etc.) of tracking down the phases specifying coherence and restoring them. This is intuitively large in many typical measurement situations. Another, related measure is the negative of the logarithm of the probability of doing so. If the probability of restoring the phases in any particular measurement situation were significant, then we would not have the necessary amount of decoherence. The correlation could not be inside the set of decohering histories. Thus, this measure of irreversibility is large. Indeed, in many circumstances where the phases are carried off to infinity or lost in photons impossible to retrieve, the probability of recovering them is truly zero and the situation perfectly irreversible—infinitely costly to reverse and with zero probability for reversal!

Defining a measurement situation solely as the existence of correlations in a quasiclassical domain, if suitable general definitions of maximality and classicity can be found, would have the advantages of clarity, economy, and generality. Measurement situations occur throughout the universe and without the necessary intervention of anything as sophisticated as an "observer." Thus, by this definition, the production of fission tracks in mica deep in the earth by the decay of a uranium nucleus leads to a measurement situation in a quasiclassical domain in which the tracks' directions decohere, whether or not these tracks are ever registered by an "observer."

XII. Complex Adaptive Systems

Our picture is of a universe that, as a consequence of a particular initial condition and of the underlying Hamiltonian, exhibits at least one quasiclassical domain made up of suitably defined maximal sets of alternative histories with as much classicity as possible. The quasiclassical domains would then be a consequence of the theory and its boundary condition, not an artifact of our construction. How do we then characterize our place as a collective of observers in the universe?

Both singly and collectively we are examples of the general class of complex adaptive systems. When they are considered within quantum mechanics as portions of the universe, making observations, we refer to such complex adaptive systems as information gathering and utilizing systems (IGUSes). The general characterization of complex adaptive systems is the subject of much ongoing research, which we cannot discuss here. From a quantum-mechanical point of view, the foremost characteristic of an IGUS is that in some form of approximation, however crude or classical, it employs the fundamental formula, with what amounts to a rudimentary theory of ρ, H, and quantum mechanics. Probabilities of interest to the IGUS include those for correlations between its memory and the external world. (Typically these are assumed to be perfect—not always such a good approximation!) The approximate fundamental formula is used to compute probabilities on the basis of present data, make predictions, control future perceptions on the basis of these predictions (i.e., exhibit behavior), acquire further data, make further predictions, and so on.

To carry on in this way, an IGUS uses probabilities for histories referring both to the future and to the past. An IGUS uses decohering sets of alternative histories and therefore performs further coarse graining on a quasiclassical domain. Naturally, its coarse graining is very much coarser than that of the quasiclassical domain since it utilizes only a few of the variables in the universe.

The reason such systems as IGUSes exist, functioning in such a fashion, is to be sought in their evolution within the universe. It seems likely that they evolved to make predictions because it is adaptive to do so.[16] The reason, therefore, for their focus on decohering variables is that these are the *only* variables for which predictions can be made. The reason for their focus on the histories of a quasiclassical domain is that these present enough regularity over time to permit the generation of models (schemata) with significant predictive power.

If there is essentially only one quasiclassical domain, then naturally the IGUS utilizes further coarse grainings of it. If there are many essentially inequivalent quasiclassical domains, then we could adopt a subjective point of view, as in some traditional discussions of quantum mechanics, and say that the IGUS "chooses" its coarse graining of histories and, therefore, "chooses" a particular quasiclassical domain, or a subset of such domains, for further coarse graining. It would be better, however, to say that the IGUS evolves to exploit a particular quasiclassical domain or set of such domains. Then IGUSes, including human beings, occupy no special place and play no preferred role in the laws of physics. They merely utilize the probabilities presented by quantum mechanics in the context of a quasiclassical domain.

XIII. Conclusions

We have sketched a program for understanding the quantum mechanics of the universe and the quantum mechanics of the laboratory, in which the notion of quasiclassical domain plays

---

[16]Perhaps, as Unruh (1986) has suggested, there are complex adaptive systems, making no use of prediction, that can function in a highly quantum-mechanical way. If this is the case, they are very different from anything we know or understand.

a central role. To carry out that program, it is important to complete the definition of a quasiclassical domain by finding the general definition for classicity. Once that is accomplished, the question of how many and what kinds of essentially inequivalent quasiclassical domains follow from ρ and H becomes a topic for serious theoretical research. So is the question of what are the general properties of IGUSes that can exist in the universe exploiting various quasiclassical domains, or the unique one if there is essentially only one.

It would be a striking and deeply important fact of the universe if among its maximal sets of decohering histories, there were one roughly equivalent group with much higher classicities than all the others. That would then be *the* quasiclassical domain, completely independent of any subjective criterion, and realized within quantum mechanics by utilizing only the initial condition of the universe and the Hamiltonian of the elementary particles.

Whether the universe exhibits one or many maximal sets of branching alternative histories with high classicities, those quasiclassical domains are the possible arenas of prediction in quantum mechanics.

It might seem at first sight that in such a picture the complementarity of quantum mechanics would be lost; in a given situation, for example, *either* a momentum *or* a coordinate could be measured, leading to different kinds of histories. We believe that impression is illusory. The histories in which an observer, as part of the universe, measures p and the histories in which that observer measures x are decohering alternatives. The important point is that the decoherent histories of a quasiclassical domain contain all possible choices that might be made by all possible observers that might exist, now, in the past, or in the future for that domain.

The Einstein–Podolsky–Rosen (EPR) or EPRB situation is no more mysterious. There, a choice of measurements, say σ_x or σ_y for a given electron, is correlated with the behavior of σ_x or σ_y for another electron because the two together are in a singlet spin state even though widely separated. Again, the two measurement situations (for σ_x and σ_y) decohere from each other, but here, in each, there is also a correlation between the information obtained about one spin and the information that can be obtained about the other. This behavior, although unfortunately called "non-local" by some authors, involves no non-locality in the ordinary sense of quantum field theory and no possibility of signaling outside the light cone. The problem with the "local realism" that Einstein would have liked is not the locality but the *realism*. Quantum mechanics describes *alternative* decohering histories and one cannot assign "reality" simultaneously to different alternatives because they are contradictory. Everett (1957) and others (DeWitt 1970) have described this situation, not incorrectly, but in a way that has confused some, by saying that the histories are all "equally real" (meaning only that quantum mechanics prefers none over another except via probabilities) and by referring to "many worlds" instead of "many histories."

We conclude that resolution of the problems of interpretation presented by quantum mechanics is not to be accomplished by further intense scrutiny of the subject as it applies to reproducible laboratory situations, but rather through an examination of the origin of the universe and its subsequent history. Quantum mechanics is best and most fundamentally understood in the context of quantum cosmology. The founders of quantum mechanics were right in pointing out that something external to the framework of wave function and Schrödinger equation is needed to interpret the theory. But it is not a postulated classical world to which quantum mechanics does not apply. Rather it is the initial condition of the universe that, together with the action function of the elementary particles and the throws of quantum dice since the beginning, explains the origin of quasiclassical domain(s) within quantum theory itself. ☙

Acknowledgments

One of us, MG-M, would like to acknowledge the great value of conversations about the meaning of quantum mechanics with Felix Villars and Richard Feynman in 1963–64 and again with Richard Feynman in 1987–88. He is also very grateful to Valentine Telegdi for discussions during 1985–86, which persuaded him to take up the subject again after twenty years. Both of us are indebted to Telegdi for further interesting conversations since 1987. We would also like to thank R. Griffiths for a useful communication and a critical reading of the manuscript and R. Penrose for a helpful discussion.

Part of this work was carried out at various times at the Institute for Theoretical Physics, Santa Barbara; the Aspen Center for Physics; the Santa Fe Institute; and the Department of Applied Mathematics and Theoretical Physics, University of Cambridge. We are grateful for the hospitality of these institutions. The work of JBH was supported in part by NSF grant PHY85-06686 and by a John Simon Guggenheim Fellowship. The work of MG-M was supported in part by the US Department of Energy under contract DE-AC-03-81ER40050 and by the Alfred P. Sloan Foundation.

Note on References

For a subject as large as this one, it would be an enormous task to cite the literature in any historically complete way. We have attempted to cite only papers that we feel will be directly useful to the points raised in the text. These are not always the earliest, nor are they always the latest. In particular, we have not attempted to review or to cite papers where similar problems are discussed from different points of view.

REFERENCES

Aharonov, Y., P. G. Bergmann, and J. L. Lebowitz. 1964. "Time Symmetry in the Quantum Process of Measurement." *Physical Review* 134 (6B): B1410–16. doi:10.1103/physrev.134. b1410.

Bohr, N. 1958. *Atomic Physics and Human Knowledge*. New York, NY: John Wiley.

Caldeira, A. O., and A. J. Leggett. 1983. "Path Integral Approach to Quantum Brownian Motion." *Physica A: Statistical Mechanics and its Applications* 121 (3): 587–616. doi:10. 1016/0378-4371(83)90013-4.

Coleman, S. 1988. "Why There Is Nothing Rather than Something: A Theory of the Cosmological Constant." *Nuclear Physics B* 310 (3–4): 643–68. doi:10.1016/0550-3213(88) 90097-1.

Cooper, L. N., and D. Van Vechten. 1969. "On the Interpretation of Measurement within the Quantum Theory." *American Journal of Physics* 37 (12): 1212–20. doi:10.1119/1.1975279.

Daneri, A., A. Loinger, and G.M. Prosperi. 1962. "Quantum Theory of Measurement and Ergodicity Conditions." *Nuclear Physics* 33:297–319. doi:10.1016/0029-5582(62)90528-x.

DeWitt, B. S. 1970. "Quantum Mechanics and Reality." *Physics Today* 23 (9): 30–35. doi:10.1063/ 1.3022331.

DeWitt, B. S., and R. N. Graham, eds. 1973. *The Many Worlds Interpretation of Quantum Mechanics*. Princeton, NJ: Princeton University Press.

Dicke, R. H. 1981. "Interaction-Free Quantum Measurements: A Paradox?" *American Journal of Physics* 49 (10): 925–30. doi:10.1119/1.12592.

Everett, H. 1957. "'Relative State' Formulation of Quantum Mechanics." *Reviews of Modern Physics* 29 (3): 454–62. doi:10.1103/revmodphys.29.454.

Farhi, E., J. Goldstone, and S. Gutmann. 1989. "How Probability Arises in Quantum Mechanics." *Annals of Physics* 192 (2): 368–82. doi:10.1016/0003-4916(89)90141-3.

Feynman, R. P., and F. L. Vernon. 1963. "The Theory of a General Quantum System Interacting with a Linear Dissipative System." *Annals of Physics* 24:118–73. doi:10.1016/0003-4916(63) 90068-x.

Finkelstein, D. 1963. "Section of Physical Sciences: The Logic of Quantum Physics." *Transactions of the New York Academy of Sciences* 25 (6 Series II): 621–37. doi:10.1111/ j.2164-0947.1963.tb01483.x.

Forster, D. 1975. *Hydrodynamic Fluctuations, Broken Symmetry, and Correlation Functions.* Reading, MA: Benjamin.

Fukuyama, T., and M. Morikawa. 1989. "Two-Dimensional Quantum Cosmology: Directions of Dynamical and Thermodynamic Arrows of Time." *Physical Review D* 39 (2): 462–69. doi:10.1103/physrevd.39.462.

Gell-Mann, M. 1963. Unpublished manuscript.

———. 1987. "Superstring Theory." *Physica Scripta* T15:202–9. doi:10.1088/0031-8949/1987/t15/ 029.

Gell-Mann, Murray. 1989. "Dick Feynman—The Guy in the Office Down the Hall." *Physics Today* 42 (2): 50–54. doi:10.1063/1.881192.

Geroch, R. 1984. "The Everett Interpretation." *Noûs* 18 (4): 617–33. doi:10.2307/2214880.

Giddings, S. B., and A. Strominger. 1988. "Loss of Incoherence and Determination of Coupling Constants in Quantum Gravity." *Nuclear Physics B* 307 (4): 854–66. doi:10.1016/0550- 3213(88)90109-5.

Graham, R. N. 1973. "The Measurement of Relative Frequency." In *The Many Worlds Interpretation of Quantum Mechanics,* edited by B. S. DeWitt and R. N. Graham. Princeton, NJ: Princeton University Press.

Griffiths, R. B. 1984. "Consistent Histories and the Interpretation of Quantum Mechanics." *Journal of Statistical Physics* 36 (1–2): 219–72. doi:10.1007/bf01015734.

Halliwell, J. J. 1988a. "A Bibliography of Papers on Quantum Cosmology." ITP preprint NSF- ITP-88-132.

———. 1988b. "Quantum Cosmology: An Introductory Review." ITP preprint NSF-ITP-88-131.

———. 1989. "Decoherence in Quantum Cosmology." *Physical Review D* 39 (10): 2912–23. doi:10.1103/physrevd.39.2912.

Hartle, J. B. 1968. "Quantum Mechanics of Individual Systems." *American Journal of Physics* 36 (8): 704–12. doi:10.1119/1.1975096.

———. 1988a. "Quantum Kinematics of Spacetime. I. Nonrelativistic Theory." *Physical Review D* 37 (10): 2818–32. doi:10.1103/physrevd.37.2818.

——. 1988b. "Quantum Kinematics of Spacetime. II. A Model Quantum Cosmology with Real Clocks." *Physical Review D* 38 (10): 2985–99. doi:10.1103/physrevd.38.2985.

——. 1989a. "Quantum Cosmology." In *Highlights in Gravitation and Cosmology*, edited by B. R. Iyer, A. Kembhavi, J. V. Narlikar, and C. V. Vishveshwara. Cambridge, UK: Cambridge University Press.

——. 1989b. "Time and Prediction in Quantum Cosmology." In *Proceedings of the 5th Marcel Grossmann Meeting on Recent Developments in General Relativity*, edited by D. G. Blair and M. J. Buckingham. Singapore: World Scientific.

——. 1990a. "Progress in Quantum Cosmology." In *Proceedings of the 12th International Conference on General Relativity and Gravitation*, edited by N. Ashby, D. F. Bartlett, and W. Wyss. Cambridge, UK: Cambridge University Press.

——. 1990b. "The Quantum Mechanics of Cosmology." In *Quantum Cosmology and Baby Universes (Proceedings of the 1989 Jerusalem Winter School in Theoretical Physics)*, edited by S. Coleman, J. B. Hartle, and T. Piran. Singapore: World Scientific.

——. 1990c. "Time and Prediction in Quantum Cosmology." In *Proceedings of the Osgood Hill Conference on the Conceptual Problems of Quantum Gravity*, edited by A. Ashtekar and J. Stachel. Boston, MA: Birkhauser.

Hartle, J. B., and S. W. Hawking. 1983. "Wave Function of the Universe." *Physical Review D* 28 (12): 2960–75. doi:10.1103/physrevd.28.2960.

Hawking, S. W. 1987. "Quantum Coherence Down the Wormhole." *Physics Letters B* 195 (3): 337–43. doi:10.1016/0370-2693(87)90028-1.

Hobson, A. 1971. *Concepts in Statistical Mechanics*. New York, NY: Gordon and Breach.

Joos, E. 1986. "Why Do We Observe a Classical Spacetime?" *Physics Letters A* 116 (1): 6–8. doi:10.1016/0375-9601(86)90345-2.

Joos, E., and H. D. Zeh. 1985. "The Emergence of Classical Properties through Interaction with the Environment." *Zeitschrift für Physik B Condensed Matter* 59 (2): 223–43. doi:10.1007/bf01725541.

Kadanoff, L. P., and P. C. Martin. 1963. "Hydrodynamic Equations and Correlation Functions." *Annals of Physics* 24:419–69. doi:10.1016/0003-4916(63)90078-2.

Kiefer, C. 1987. "Continuous Measurement of Mini-Superspace Variables by Higher Multipoles." *Classical and Quantum Gravity* 4 (5): 1369–82. doi:10.1088/0264-9381/4/5/031.

Landau, L., and E. Lifshitz. 1958. *Quantum Mechanics*. London, UK: Pergamon.

Lebowitz, J. L. 1986. "Microscopic Origin of Hydrodynamic Equations: Derivation and Consequences." *Physica A: Statistical Mechanics and its Applications* 140 (1–2): 232–39. doi:10.1016/0378-4371(86)90227-x.

Lloyd, S. Private communication.

London, F., and E. Bauer. 1939. *La théorie de l'observation en mécanique quantique.* Paris, France: Hermann.

Mukhanov, V. F. 1985. "On the Many-Worlds Interpretation of Quantum Theory." In *Proceedings of the Third Seminar on Quantum Gravity,* edited by M. A. Markov, V. A. Berezin, and V. P. Frolov. Singapore: World Scientific.

Omnès, R. 1988a. "Logical Reformulation of Quantum Mechanics. I. Foundations." *Journal of Statistical Physics* 53 (3–4): 893–932. doi:10.1007/bf01014230.

———. 1988b. "Logical Reformulation of Quantum Mechanics. II. Interferences and the Einstein-Podolsky-Rosen Experiment." *Journal of Statistical Physics* 53 (3–4): 933–55. doi:10.1007/bf01014231.

———. 1988c. "Logical Reformulation of Quantum Mechanics. III. Classical Limit and Irreversibility." *Journal of Statistical Physics* 53 (3–4): 957–75. doi:10.1007/bf01014232.

Peierls, R. B. 1985. "Observations in Quantum Mechanics and the Collapse of the Wave Function." In *Symposium on the Foundations of Modern Physics,* edited by P. Lahti and P. Mittelstaedt. Singapore: World Scientific.

Renninger, M. 1960. "Messungen ohne Störung des Meßobjekts." *Zeitschrift für Physik* 158 (4): 417–21. doi:10.1007/bf01327019.

Rosenkrantz, R. D., ed. 1983. *E. T. Jaynes: Papers on Probability, Statistics, and Statistical Physics.* Dordrecht, Netherlands: D. Reidel.

Unruh, W. G. 1986. "Quantum Measurement." In *New Techniques and Ideas in Quantum Measurement Theory,* edited by D. M. Greenberger, *Annals of the New York Academy of Sciences* 480:242–49. doi:10.1111/j.1749-6632.1986.tb12427.x.

Vilenkin, A. 1986. "Boundary Conditions in Quantum Cosmology." *Physical Review D* 33 (12): 3560–69. doi:10.1103/physrevd.33.3560.

Wheeler, J. A. 1957. "Assessment of Everett's "Relative State" Formulation of Quantum Theory." *Reviews of Modern Physics* 29 (3): 463–65. doi:10.1103/revmodphys.29.463.

Wigner, E. P. 1963. "The Problem of Measurement." *American Journal of Physics* 31 (1): 6–15. doi:10.1119/1.1969254.

Zeh, H. D. 1970. "On the Interpretation of Measurement in Quantum Theory." *Foundations of Physics* 1 (1): 69–76. doi:10.1007/bf00708656.

———. 1986. "Emergence of Classical Time from a Universal Wavefunction." *Physics Letters A* 116 (1): 9–12. doi:10.1016/0375-9601(86)90346-4.

Zurek, W. H. 1981. "Pointer Basis of Quantum Apparatus: Into What Mixture Does the Wave Packet Collapse?" *Physical Review D* 24 (6): 1516–25. doi:10.1103/physrevd.24.1516.

———. 1982. "Environment-Induced Superselection Rules." *Physical Review D* 26 (8): 1862–80. doi:10.1103/physrevd.26.1862.

———. 1984. "The Reduction of the Wavepacket: How Long Does It Take?" In *Frontiers of Non-Equilibrium Quantum Statistical Physics,* edited by G. Moore and M. Scully. New York, NY: Plenum Press.

———. 1989. "Algorithmic Randomness and Physical Entropy." *Physical Review A* 40 (8): 4731–51. doi:10.1103/physreva.40.4731.

6 ⊐

INFORMATION DISSIPATION IN QUANTUM COSMOLOGY AND THE EMERGENCE OF CLASSICAL SPACETIME

Jonathan J. Halliwell, Institute for Theoretical Physics,

University of California, Santa Barbara

We discuss the manner in which the gravitational field becomes classical in quantum cosmology—quantum gravity applied to closed cosmologies. We argue that there are at least two steps involved. First, the quantum state of the gravitational field must be strongly peaked about a set of classical configurations. Second, these configurations must have negligible interference with each other. This second step involves decoherence—destruction of the off-diagonal terms in the density matrix, representing interference. This may be achieved by dissipating information about correlations into an environment. Although the entire universe, by definition, has no environment, it may be split up into subsystems and one or more subsystems may be regarded as an environment for the others. In particular, matter modes may be used as an environment for the gravitational field. We show, in a simple homogeneous isotropic model, that the density matrix of the universe is decohered by the long-wavelength modes of an inhomogeneous massless scalar field. We also show, using decoherence arguments, that the WKB component of the wave function of the universe which represents expanding universes has negligible interference with the collapsing component. This justifies the usual assumption that they may be treated separately. We discuss the role of cosmological boundary conditions. The fact that we observe a classical spacetime today seems to depend on them crucially.

1. Introduction

The point of this chapter is to discuss some recent work on the application of a body of ideas normally used in quantum measurement theory to quantum cosmology. The question that

I will address is the following: How, in a quantum theory of gravity as applied to closed cosmological systems, i.e., in quantum cosmology, does the gravitational field become classical? The possible answer to this question that I will discuss involves decoherence of the density matrix of the universe. This necessarily involves the *dissipation of information*, making contact with the information theme of this meeting. But before proceeding to quantum cosmology, we begin by discussing the emergence of classical behavior in some more down-to-earth quantum systems.

It is one of the undeniable facts of our experience that the world about us is described by classical laws to a very high degree of accuracy. In classical mechanics, a system may be assigned a quite definite state and its evolution is described in a deterministic manner—given the state of the system at a particular time, one can predict its state at a later time with certainty. And yet, it is believed that the world is fundamentally quantum mechanical in nature. Phenomena on all scales up to and including the entire universe are supposedly described by quantum mechanics. In quantum mechanics, because superpositions of interfering states are permissible, it is generally not possible to say that a system is in a definite state. Moreover, evolution is not deterministic but probabilistic— given the state of the system at a particular time, one can calculate only the probability of finding it in another state at a later time.

If quantum theory is to be reconciled with our classical experience, it is clearly essential to understand the sense in which, and the extent to which, quantum mechanics reproduces the effects of classical mechanics. This is an issue that assumes particular importance in the quantum theory of measurement (Wheeler and Zurek 1983). There, one describes the measuring apparatus in quantum-mechanical terms; yet all such apparata behave in a distinctly classical manner when the experimenter's eye reads the meter.

Early-universe cosmology provides another class of situations in which the emergence of classical behavior from quantum mechanics is a process of particular interest. In the inflationary universe scenario, for example, the classical density fluctuations required for galaxy formation supposedly originate in the quantum fluctuations of a scalar field, hugely amplified by inflation (Guth and Pi 1982; Hawking 1982). This is, in a sense, an extreme example of a quantum measurement process, in that the large-scale structure of the universe that we see today is a meter which has permanently recorded the quantum state of the scalar field at early times. The manner in which this quantum-to-classical transition comes about has been discussed by numerous authors.[1] A more fundamental situation of interest, and the one with which we are primarily concerned, is quantum cosmology in which one attempts to apply quantum mechanics to closed cosmologies. Since this involves quantizing the gravitational field, one of the goals of this endeavor should surely be to predict the conditions under which the gravitational field may be regarded as classical. And at a humbler level, one can ask why everyday objects such as tables and chairs behave classically when they are really described by quantum mechanics.

The point of view which we will take is the following: There are at least two requirements that must be satisfied before a system may be regarded as classical. The first requirement is that the wave function of the system, or some distribution constructed from the wave function, should be

-679-

---

[1]See Guth and Pi (1985), Halliwell (1987b), Lyth (1985), Sakagami (1987), and Sasaki (1986).

strongly peaked about a classical configuration or a set of classical configurations. This requirement would be satisfied, for example, if the wave function were a coherent state, or a superposition of coherent states. Even though this requirement means that the quantum state may be peaked about distinct macroscopic configurations, it does not, however, rule out the possibility of interference between them. The second requirement, therefore, is that the interference between distinct macroscopic states is exceedingly small. This involves the notion of *decoherence*—destruction of the off-diagonal terms in the density matrix, which represent interference.

In the next section, we will make these ideas more precise in a simple example from ordinary quantum mechanics. In the subsequent sections, we will go on to discuss how these ideas are applied to quantum cosmology. A much longer, but slightly different, account of this work may be found in Halliwell (1989).

2. The Emergence of Classical Behavior in Quantum Mechanics

Consider a single-particle system S which starts out in a state $|\Psi(0)\rangle$ and after a time t finds itself in a superposition of well-separated coherent states:

$$|\Psi(t)\rangle = \sum_n c_n |x_n(t)\rangle. \qquad (2.1)$$

In the configuration space representation, the coherent states $|x_n(t)\rangle$ are given by

$$\langle x|x_n(t)\rangle = \exp(ip_n x) \exp\left(-\frac{(x - x_n(t))^2}{\sigma^2}\right). \qquad (2.2)$$

They are Gaussian wave packets strongly peaked about the classical trajectories $x_n(t)$. One might therefore be tempted to say that the system has become classical, and that the particle will be following one of the trajectories $x_n(t)$ with probability $|c_n|^2$. The problem, however, is that if the wave packets met up at some stage in the future, then they would interfere constructively. One could not, therefore, say that the particle is following a definite trajectory.

The problem is highlighted when one writes down the pure-state density matrix corresponding to the state in equation (2.1). It is

$$\rho_{\text{pure}}(t) = |\Psi(t)\rangle\langle\Psi(t)| = \sum_{n,m} c_m^* c_n \, |x_n(t)\rangle\langle x_m(t)|. \qquad (2.3)$$

It involves non-zero, off-diagonal terms which represent interference between different trajectories. We are seeking to maintain, however, description of the system by a classical ensemble, of the type encountered in statistical mechanics, in which one finds the particle to be following the trajectory $x_n(t)$ with probability $|c_n|^2$. Such a situation could only be described by the mixed-state density matrix

$$\rho_{\text{mixed}} = \sum_{n} |c_n|^2 \, |x_n(t)\rangle\langle x_n(t)|. \qquad (2.4)$$

This differs from equation (2.3) by the presence of off-diagonal terms. It is only when these terms may be neglected that we may say that the particle is following a definite trajectory.

There is no way that under unitary Schrödinger evolution the pure-state density matrix in equation (2.3) will evolve into the mixed-state density matrix in equation (2.4). How, then, may the interference terms be suppressed? The resolution of this apparent difficulty comes from the recognition that no macroscopic system can realistically be considered as closed and isolated from the rest of the world around it. Laboratory measuring apparata interact with surrounding air molecules; even intergalactic gas molecules are not isolated because they interact with the microwave background. Let us refer to the rest of the world as "the environment," $\mathcal{E}$. Then it can be argued that it is the inescapable interaction with the environment which leads to a continuous "measuring" or "monitoring" of a macroscopic system and it is this that causes the interference terms to become very small. This is decoherence.

Let us study this in more detail. Consider the system $\mathcal{S}$ considered above, but now take into account also the states $\{|\mathcal{E}_n\rangle\}$ of the environment $\mathcal{E}$. Let the initial state of the total system $\mathcal{SE}$ be

$$|\Phi(0)\rangle = |\Psi(0)\rangle\,|\mathcal{E}_0\rangle, \qquad (2.5)$$

where $|\mathcal{E}_0\rangle$ is the initial state of the environment. After time t, this will evolve into a state of the form

$$|\Phi(t)\rangle = \sum_n c_n\,|x_n(t)\rangle\,|\mathcal{E}_n\rangle. \qquad (2.6)$$

The coherent states of the system, $|x_n(t)\rangle$, thus become correlated with the environment states $|\mathcal{E}_n\rangle$. The point, however, is that one is not interested in the state of the environment. This is traced out in the calculation of any quantities of interest. The object of particular relevance, therefore, is the reduced or coarse-grained density matrix, obtained by tracing over the environment states:

$$\tilde{\rho} = \mathrm{Tr}_{\mathcal{E}}\,|\Phi(t)\rangle\langle\Phi(t)| = \sum_{n,m} c_n c_m^*\,\langle\mathcal{E}_m|\mathcal{E}_n\rangle\,|x_n(t)\rangle\langle x_m(t)|.$$

$$(2.7)$$

The density matrix $|\Phi(t)\rangle\langle\Phi(t)|$ of the total system evolves unitarily, of course. The reduced density matrix in equation (2.7), however, does not. It therefore holds the possibility of evolving an initially pure state to a final mixed state. In particular, if, as can be the case, the inner products $\langle\mathcal{E}_m|\mathcal{E}_n\rangle$ are very small when $n \neq m$, then equation (2.7) will be indistinguishable from the mixed-state density matrix in equation (2.4).

One may now say that the environment has caused the density matrix to decohere—it has permitted the interfering set of macroscopic configurations to resolve into a non-interfering ensemble of states, as used in classical statistical mechanics. Or to put it another way, the environment has "collapsed the wave function" of the system. Or yet another form of words is to say that the environment "induces a superselection rule" which prevents superpositions of distinct macroscopic states from being observed. Note that the *loss of information* is an important aspect of the process. Classical behavior thus emerges only when information about correlations is dissipated into the environment.

~ 683 ~

This general body of ideas has been discussed by many people, including Gell-Mann, Hartle, and Telegdi, Griffiths (1984, 1987), Joos and Zeh (1985), Omnès (1988), Peres (1980), Unruh and Zurek (1989), Wigner (1982), Zeh (1970), and Zurek (1981, 1982, 1986, 1989).

3. Quantum Cosmology

We now apply the ideas introduced in the previous section to quantum cosmology. This subject began life in the 1960s, with the seminal works of DeWitt (1967), Misner (1969a, 1969b, 1970, 1972), and Wheeler (1964, 1968). More recently, it has been revitalized primarily by Hartle and Hawking (1983) and by Vilenkin (1982, 1983, 1984, 1985, 1986). Some review articles are those by Hartle (1985, 1987) and Halliwell (1988).

The object is that one applies ideas from an as-yet-incomplete quantum theory of gravity to closed cosmological models. One imagines that the four-dimensional spacetime is sliced up into three-surfaces, and one concentrates on the variables defined on the three-surfaces which describe the configuration of the gravitational and matter fields. These are the three-metric h_{ij} and the matter field, which we take to be a scalar field Φ. The quantum state of the system is then represented by a wave functional $\Psi[h_{ij}, \Phi]$, a functional of the metric and scalar field configurations. For rather fundamental reasons, the wave functional does not depend on time explicitly. Loosely speaking, information about time is already contained in the variables h_{ij} and Φ. Because it does not have an explicit time label, Ψ obeys not a time-dependent Schrödinger equation but a zero-energy equation of the form

$$(H_g + H_m)\Psi = 0, \qquad (3.1)$$

where H_g and H_m are, respectively, the gravitational and matter Hamiltonians.

Suppose one solves the Wheeler–DeWitt equation subject to certain boundary conditions. Then one finds that, in certain regions, there exist approximate Wentzel–Kramers–Brillouin (WKB) solutions of the form

$$\Psi[h_{ij}] = \exp(iS[h_{ij}])\Psi_m[h_{ij}, \Phi]. \qquad (3.2)$$

Here $S[H_{ij}]$ is a rapidly varying phase and satisfies the Einstein–Hamilton–Jacobi equation. The e^{iS} part of the wave function thus indicates that the wave function corresponds to a set of classical solutions to the Einstein equation with Hamilton–Jacobi function S. More precisely, one may show that equation (3.2) is strongly peaked about the set of solutions satisfying the first integral

$$\pi^{ij} = \frac{\delta S}{\delta h_{ij}}, \qquad (3.3)$$

where π^{ij} is the momentum conjugate to h_{ij} (Halliwell 1987a). The wave function in equation (3.2) is therefore analogous[2] to the sum of coherent states in equation (2.1). The wave function $\Psi_m[h_{ij}, \Phi]$ is a slowly varying function of the three-metric. It describes quantum field theory for the scalar field Φ on the gravitational background h_{ij}.

So the first requirement for the emergence of classical behavior is satisfied by the solution in equation (3.2)—the wave function is peaked about a set of classical solutions. But what about the second requirement, decoherence? Let us apply the ideas from the previous sections and introduce an environment which continually monitors the metric. One meets with an immediate difficulty. The entire universe has no environment. It is not an open system, but a closed one; in fact, it is the only genuinely closed system we know of. The point, however, is that one is never interested in measuring more than a small fraction of the potentially observable features of the universe. One may therefore regard just some of the variables describing the universe as the observed system and the rest as environment. The latter are traced out in the density matrix. In this way, some—but certainly not all—of the variables describing the universe may become classical.

Which variables do we take to be the environment? There is, in general, no obvious natural choice. However, here we are interested in understanding how the gravitational field becomes classical, so it is perhaps appropriate to regard the matter modes as an environment for the metric. With this choice, the reduced density matrix corresponding to the wave

---

[2]It is actually rather difficult to construct the analog of coherent states in quantum cosmology. See, however, Kiefer (1988).

function in equation (3.2) is

$$\tilde{\rho}(h'_{ij}, h_{ij}) = \int \mathcal{D}\Phi \Psi^*[h'_{ij}, \Phi] \Psi[h_{ij}, \Phi]. \qquad (3.4)$$

The object is to show that this is small for $h'_{ij} \neq h_{ij}$. It is very difficult to offer general arguments as to the extent to which this is the case, but one can see it for particular models. Numerous models have been considered in the literature—for example, Fukuyama and Morikawa (1988), Halliwell (1989), Kiefer (1987), Mellor and Moss (1988), Morikawa (1988), Padmanabhan (1989), and Zeh (1986, 1988).

For definiteness, let us briefly consider one particular model (Halliwell 1989). Suppose we restrict the metric to be of the Robertson–Walker type

$$ds^2 = -dt^2 + a^2(t)\, d\Omega_3^2, \qquad (3.5)$$

where $d\Omega_3^2$ is the metric on the three-sphere. Then the gravitational field is described solely by the scale factor a. Let us take the only source to be a cosmological constant Λ. One may show that the wave function for this model is of the form in equation (3.2), and the e^{iS} part indicates that it is peaked about classical solutions of the form

$$a(t) = \frac{1}{H} \cosh(Ht), \qquad (3.6)$$

where $H^2 = \Lambda/3$. This is de Sitter space. Most models that have been considered in the literature use the full infinite number of modes of the scalar field as the environment. However, this leads to certain technical complications, so here we will do something simpler. The de Sitter solutions have a horizon size $a = H^{-1}$. One may separate the scalar field modes into long (l) or short (s) wavelength modes, $\Phi = \Phi_l + \Phi_s$, depending on whether their wavelength is, respectively, greater or less than the horizon size. The number of modes outside the horizon is actually finite; moreover, they are not observable, so it seems reasonable to consider them as the environment. With this choice, and with a particular choice for the quantum state of the scalar field, one finds that the reduced density matrix is

$$\tilde{\rho}(a, a') \approx \exp\left[-\frac{(a - a')^2}{\sigma^2}\right], \qquad (3.7)$$

where the coherence width σ is given by $\sigma = 1/H^3 a$. It follows that $\tilde{\rho}$ diagonalizes for $\sigma \gg a$, i.e., for $Ha \gg 1$. This means that the interference becomes negligible when the scale factor is much greater than the minimum size of de Sitter space.

4. The Interference between Expanding and Collapsing Universes

One may go further with this approach. The Wheeler–DeWitt operator is real. This means that if Ψ is a solution, then its complex conjugate is a solution also. In particular, there is a purely real solution

$$\Psi(a, \Phi) = e^{iS(a)}\Psi_m(a, \Phi) + e^{-iS(a)}\Psi_m(a, \Phi) = \Psi_{(+)} + \Psi_{(-)}. \qquad (4.1)$$

The Hartle–Hawking "no-boundary" boundary condition proposal picks out a wave function of this type (Hartle and Hawking 1983). If the first term is regarded as corresponding to a set of expanding solutions, then the second corresponds to a set of collapsing solutions.[3] Because each of these WKB solutions corresponds to very distinct macroscopic states, one would hope to show that the interference between them is negligible.

---

[3]Because there is no explicit time label, one cannot say which of the two solutions corresponds to collapsing and which corresponds to expanding— one can only make relative statements. I am grateful to H. D. Zeh for emphasizing this point to me.

This is indeed possible (Halliwell 1989). Following the above approach, and again using Φ_l as the environment, one finds that the part of the reduced density matrix for equation (4.1) corresponding to the interference between expanding and collapsing solutions is

$$\tilde{\rho}_{(+-)} = \mathrm{Tr}_{\Phi_l}\left[\Psi^*_{(+)}(a',\Phi_l)\Psi_{(-)}(a,\Phi_l)\right] \approx \exp\left[-\frac{(a+a')^2}{\sigma^2}\right].$$

(4.2)

Equation (4.2) differs from equation (3.7) in one crucial respect, namely in the sign between a and a'. This has the consequence that $\tilde{\rho}_{(+-)}$ is *always* very small, even when $a = a'$. The interference between expanding and collapsing components of the wave function may therefore be neglected.

5. Dependence on Boundary Conditions

We have seen that in quantum cosmology, spacetime may be become classical when (i) the wave function is peaked about sets of classical configurations and (ii) the interference between these configurations is destroyed through interaction with matter variables. To what extent does the emergence of classical behavior depend on boundary or initial conditions? Boundary conditions enter in two ways. The wave function of the system in a given region may be either exponential or oscillatory, depending to some extent on the boundary conditions. It is only when the wave function is oscillatory that it is peaked about a set of classical solutions; thus, the boundary conditions determine whether or not the wave function is peaked about sets of classical configurations.

The second way in which boundary conditions enter is through the quantum state of the environment. The coherence width σ will depend, possibly quite crucially, on the quantum state of the environment. This, in turn, is determined by the cosmological boundary conditions; thus, the boundary conditions will control the extent to which distinct macroscopic states decohere.

These considerations suggest that the fact that the present-day universe is described so well by classical laws is a consequence of a law of initial conditions, as has previously been suggested by Gell-Mann, Hartle, and Telegdi and by Hartle (1988a, 1988b).

~689~

6. Does the Entire Universe Really Have No Environment?

In the "Quantum Cosmology" section, it was stated that the entire universe has no environment and, for that reason, one has to split it into subsystems and regard one as an environment for the rest. This is certainly the case for conventional quantum cosmology. However, recent developments yield new perspectives on this issue. In quantum cosmology, one normally thinks of the spatial extent of universe as being represented by a single, connected three-surface. However, it has recently been suggested that it may also have a large number of small disconnected components, referred to as "baby universes."[4] In a Euclidean path integral, these baby universes are connected to the "parent universe" by wormholes. The picture one has, therefore, is of a large parent universe in a dilute gas of baby universes. The original motivation for studying this scenario is that the baby universes lead to an effective modification of the fundamental coupling constants, possibly leading to a prediction of their values. However, it is clear that baby universes could also be of value in connection with the issue studied here, namely the emergence of classical

---

[4]See Coleman (1988), Giddings and Strominger (1988a, 1988b), Hawking (1987, 1988), and Hawking and Laflamme (1988).

behavior for macroscopic systems. In particular, a possibility which naturally suggests itself is to use the baby universes as an environment to decohere the density matrix. First steps in this direction have been taken by Ellis, Mohanty, and Nanopoulos (1989). They estimated that, although the baby universes have negligible effect for single particles, they very effectively decohere the density matrix of a macroscopic body with Avogadro's number of particles. *

Acknowledgments

I would like to thank Jim Hartle, Raymond Laflamme, Seth Lloyd, Jorma Louko, Ian Moss, Don Page, and H. Dieter Zeh for useful conversations. I am particularly grateful to Wojciech Zurek for many very enlightening discussions on decoherence. I would also like to thank Wojciech for organizing such an interesting and successful meeting.

REFERENCES

Coleman, S. 1988. "Why There Is Nothing Rather than Something: A Theory of the Cosmological Constant." *Nuclear Physics B* 310 (3–4): 643–68. doi:10.1016/0550-3213(88)90097-1.

DeWitt, B. S. 1967. "Quantum Theory of Gravity. I. The Canonical Theory." *Physical Review* 160 (5): 1113–48. doi:10.1103/physrev.160.1113.

Ellis, J., S. Mohanty, and D. V. Nanopoulos. 1989. "Quantum Gravity and the Collapse of the Wavefunction." *Physics Letters B* 221 (2): 113–19. doi:10.1016/0370-2693(89)91482-2.

Fukuyama, T., and M. Morikawa. 1988. *Time Dependence of Coleman-Hawking Mechanism for the Vanishing of Cosmological Constant*. Kyoto University preprint KUNS-951.

Gell-Mann, M., J. B. Hartle, and V. Telegdi. Work in progress.

Giddings, S. B., and A. Strominger. 1988a. "Axion-Induced Topology Change in Quantum Gravity and String Theory." *Nuclear Physics B* 306 (4): 890–907. doi:10.1016/0550-3213(88)90446-4.

———. 1988b. "Loss of Incoherence and Determination of Coupling Constants in Quantum Gravity." *Nuclear Physics B* 307 (4): 854–66. doi:10.1016/0550-3213(88)90109-5.

Griffiths, R. B. 1984. "Consistent Histories and the Interpretation of Quantum Mechanics." *Journal of Statistical Physics* 36 (1–2): 219–72. doi:10.1007/bf01015734.

———. 1987. "Correlations in Separated Quantum Systems: A Consistent History Analysis of the EPR Problem." *American Journal of Physics* 55 (1): 11–17. doi:10.1119/1.14965.

Guth, A. H., and S.-Y. Pi. 1982. "Fluctuations in the New Inflationary Universe." *Physical Review Letters* 49 (15): 1110–13. doi:10.1103/physrevlett.49.1110.

———. 1985. "Quantum Mechanics of the Scalar Field in the New Inflationary Universe." *Physical Review D* 32 (8): 1899–920. doi:10.1103/physrevd.32.1899.

Halliwell, J. J. 1987a. "Correlations in the Wave Function of the Universe." *Physical Review D* 36 (12): 3626–40. doi:10.1103/physrevd.36.3626.

———. 1987b. "Scalar Fields in Cosmology with an Exponential Potential." *Physics Letters B* 185 (3–4): 341–44. doi:10.1016/0370-2693(87)91011-2.

———. 1988. *Quantum Cosmology: An Introductory Review*. Santa Barbara Institute for Theoretical Physics preprint NSF-ITP-88-131. An extensive list of papers on quantum cosmology may be found in J. J. Halliwell, ITP preprint NSF-ITP-88-132, 1988.

———. 1989. "Decoherence in Quantum Cosmology." *Physical Review D* 39 (10): 2912–23. doi:10.1103/physrevd.39.2912.

Hartle, J. B. 1985. "Quantum Cosmology." In *High Energy Physics 1985: Proceedings of the Yale Summer School,* edited by M. J. Bowick and F. Gursey. Singapore: World Scientific.

———. 1987. "Prediction in Quantum Cosmology." In *Gravitation in Astrophysics: Cargèse 1986,* edited by B. Carter and J. B. Hartle. New York, NY: Plenum Press.

———. 1988a. "Quantum Kinematics of Spacetime. I. Nonrelativistic Theory." *Physical Review D* 37 (10): 2818–32. doi:10.1103/physrevd.37.2818.

———. 1988b. "Quantum Kinematics of Spacetime. II. A Model Quantum Cosmology with Real Clocks." *Physical Review D* 38 (10): 2985–99. doi:http://dx.doi.org/10.1103/physrevd.38.2985.

Hartle, J. B., and S. W. Hawking. 1983. "Wave Function of the Universe." *Physical Review D* 28 (12): 2960–75. doi:10.1103/physrevd.28.2960.

Hawking, S. W. 1982. "The Development of Irregularities in a Single Bubble Inflationary Universe." *Physics Letters B* 115 (4): 295–97. doi:10.1016/0370-2693(82)90373-2.

Hawking, S. W. 1987. "Quantum Coherence Down the Wormhole." *Physics Letters B* 195 (3): 337–43. doi:10.1016/0370-2693(87)90028-1.

———. 1988. "Wormholes in Spacetime." *Physical Review D* 37 (4): 904–10. doi:10 . 1103 / physrevd.37.904.

Hawking, S. W., and R. Laflamme. 1988. "Baby Universes and the Non-Renormalizability of Gravity." *Physics Letters B* 209 (1): 39–41. doi:10.1016/0370-2693(88)91825-4.

Joos, E., and H. D. Zeh. 1985. "The Emergence of Classical Properties through Interaction with the Environment." *Zeitschrift für Physik B Condensed Matter* 59 (2): 223–43. doi:10.1007/bf01725541.

Kiefer, C. 1987. "Continuous Measurement of Mini-Superspace Variables by Higher Multipoles." *Classical and Quantum Gravity* 4 (5): 1369–82. doi:10.1088/0264-9381/4/5/031.

———. 1988. "Wave Packets in Minisuperspace." *Physical Review D* 38 (6): 1761–72. doi:10.1103/physrevd.38.1761.

Lyth, D. H. 1985. "Large-Scale Energy-Density Perturbations and Inflation." *Physical Review D* 31 (8): 1792–98. doi:10.1103/physrevd.31.1792.

Mellor, F., and I. G. Moss. 1988. Newcastle University preprint.

Misner, C. W. 1969a. "Mixmaster Universe." *Physical Review Letters* 22 (20): 1071–74. doi:10.1103/physrevlett.22.1071.

———. 1969b. "Quantum Cosmology. I." *Physical Review* 186 (5): 1319–27. doi:http://dx.doi.org/10.1103/physrev.186.1319.

———. 1970. "Classical and Quantum Dynamics of a Closed Universe." In *Relativity*, edited by M. Carmeli, S. I. Fickler, and L. Witten. New York, NY: Plenum Press.

———. 1972. "Minisuperspace." In *Magic without Magic: John Archibald Wheeler: A Collection of Essays in Honor of His 60th Birthday*, edited by J. Klauder. San Francisco, CA: W. H. Freeman.

Morikawa, M. 1988. *Evolution of the Cosmic Density Matrix*. Kyoto University preprint KUNS-923.

Omnès, Roland. 1988. "Logical Reformulation of Quantum Mechanics. I, II, and III." *Journal of Statistical Physics* 53 (3–4): 893–932, 933–56, 933–56. doi:10.1007/bf01014230.

Padmanabhan, T. 1989. "Decoherence in the Density Matrix Describing Quantum Three-Geometries and the Emergence of Classical Spacetime." *Physical Review D* 39 (10): 2924–32. doi:10.1103/physrevd.39.2924.

Peres, A. 1980. "Zeno Paradox in Quantum Theory." *American Journal of Physics* 48 (11): 931–32. doi:10.119/1.12204.

Sakagami, M. 1987. Hiroshima University preprint RRK 87-5.

Sasaki, M. 1986. "Large Scale Quantum Fluctuations in the Inflationary Universe." *Progress of Theoretical Physics* 76 (5): 1036–46. doi:10.1143/ptp.76.1036.

Unruh, W. G., and W. H. Zurek. 1989. "Reduction of a Wave Packet in Quantum Brownian Motion." *Physical Review D* 40 (4): 1071–94. doi:10.1103/physrevd.40.1071.

Vilenkin, A. 1982. "Creation of Universes from Nothing." *Physics Letters B* 117 (1–2): 25–28. doi:10.1016/0370-2693(82)90866-8.

———. 1983. "Birth of Inflationary Universes." *Physical Review D* 27 (12): 2848–55. doi:10.1103/physrevd.27.2848.

———. 1984. "Quantum Creation of Universes." *Physical Review D* 30 (2): 509–11. doi:10.1103/physrevd.30.509.

———. 1985. "Quantum Origin of the Universe." *Nuclear Physics B* 252:141–52. doi:10.1016/0550-3213(85)90430-4.

———. 1986. "Boundary Conditions in Quantum Cosmology." *Physical Review D* 33 (12): 3560–69. doi:10.1103/physrevd.33.3560.

Wheeler, J. A. 1964. "Geometrodynamics and the Issue of the Final State." In *Relativity, Groups, and Topology: 1963 Les Houches Lectures*, edited by C. DeWitt and B. DeWitt. New York, NY: Gordon and Breach.

———. 1968. "Super Space and the Nature of Quantum Geometrodynamics." In *Batelles Rencontres*, edited by B. S. DeWitt and J. A. Wheeler. New York, NY: Benjamin.

Wheeler, J. A., and W. H. Zurek. 1983. *Quantum Theory and Measurement*. Princeton, NJ: Princeton University Press.

Wigner, E. P. 1982. "Review of the Quantum Mechanical Measurement Problem." In *Quantum Optics, Experimental Gravitation, and Measurement Theory*, edited by G. T. Moore and M. O. Scully. New York, NY: Plenum Press.

Zeh, H. D. 1970. "On the Interpretation of Measurement in Quantum Theory." *Foundations of Physics* 1 (1): 69–76. doi:10.1007/bf00708656.

———. 1986. "Emergence of Classical Time from a Universal Wavefunction." *Physics Letters A* 116 (1): 9–12. doi:10.1016/0375-9601(86)90346-4.

———. 1988. "Time in Quantum Gravity." *Physics Letters A* 126 (5–6): 311–17. doi:10.1016/0375-9601(88)90842-0.

Zurek, W. H. 1981. "Pointer Basis of Quantum Apparatus: Into What Mixture Does the Wave Packet Collapse?" *Physical Review D* 24 (6): 1516–25. doi:10.1103/physrevd.24.1516.

———. 1982. "Environment-Induced Superselection Rules." *Physical Review D* 26 (8): 1862–80. doi:10.1103/physrevd.26.1862.

Zurek, W. H. 1986. "Pointer Basis of Quantum Apparatus: Into What Mixture Does the Wave Packet Collapse?" In *Frontiers of Nonequilibrium Statistical Physics*, edited by G. T. Moore and M. O. Scully. New York, NY: Plenum Press.

————. 1989. In *Proceedings of the Osgood Hill Conference on Conceptual Problems in Quantum Gravity*, edited by A. Ashtekar and J. Stachel. Boston, MA: Birkhäuser.

꒦ꕥ

THE QUANTUM MECHANICS OF
SELF-MEASUREMENT

David Z. Albert, Columbia University

Introduction

Let me start off by telling you a science-fiction story that is essentially in the tradition of curious stories about quantum mechanics, like the story about Schrödinger's cat and the story about Wigner's friend. Those stories both begin with the assumption that every physical system in the world (not merely subatomic particles, but measuring instruments and tables and chairs and cats and people and oceans and stars, too) is a quantum-mechanical system, that all such systems evolve entirely in accordance with the linear quantum-mechanical equations of motion, and that every self-adjoint local operator of such systems can, at least in principle, be measured. Those are the rules of the game we're going to play here; and what I want to tell you about is a move which is possible in this game, but which hasn't been considered before.

The old stories of Schrödinger's cat and Wigner's friend end at a point where (in the first case) the cat is in a superposition of states, one in which it is alive and the other in which it is dead; or where (in the second case) the friend is in a superposition of states that entail various mutually exclusive beliefs about the result of some given experiment. Suppose, for example, that Wigner's friend carries out a measurement of the y-spin of a spin $-\frac{1}{2}$ particle p that is initially prepared in the state $[\sigma_2 = +\frac{1}{2}\rangle_p$. He carries out the measurement by means of a measuring device that interacts with p and that he subsequently looks at in order to ascertain the result of the measurement. The end of that story looks like this:

$$|\alpha\rangle = \frac{1}{\sqrt{2}}\left[[\text{Believes that } \sigma_y = +\tfrac{1}{2}\rangle_{\text{Friend}}\right.$$

$$\left.\cdot [\text{Shows that } \sigma_y = +\tfrac{1}{2}\rangle_{\text{Measuring Device}} \cdot [\sigma_y = +\tfrac{1}{2}\rangle_p\right]$$

$$+ \left[[\text{Believes that } \sigma_y = -\tfrac{1}{2}\rangle_{\text{Friend}}\right.$$

$$\left.\cdot [\text{Shows that } \sigma_y = -\tfrac{1}{2}\rangle_{\text{Measuring Device}} \cdot [\sigma_y = -\tfrac{1}{2}\rangle_p\right].$$

(The phrase "Believes that $\sigma_y = \tfrac{1}{2}$," of course, doesn't completely specify the quantum state of Wigner's friend's very complicated brain. But the many other degrees of freedom of that system—those, for example, that specify what sort of ice cream Wigner's friend prefers—simply don't concern us here, and so, for the moment, we'll ignore them.) Now, such endings as this are usually judged to be so bizarre, and so blatantly to contradict daily experience, as to invalidate the assumption that gives rise to these stories. That is, these stories are usually judged to imply that there must be physical processes in the world that *cannot* be described by linear equations of motion, processes like the collapse of the wave function.

There are, on the other hand, as everybody knows, a number of ways of attempting to *deny* this judgment; there are a number of ways of attempting to suppose that *this* is genuinely the way things *are* at the end of a measuring process, and that this state somehow manages to *appear* to Wigner's friend or to *count* for Wigner's friend as either a case of believing that $\sigma_y = +\tfrac{1}{2}$ or a case of believing that $\sigma_y = -\tfrac{1}{2}$.

One of these attempts goes back to Hugh Everett, and has come to be called the many-worlds interpretation of quantum mechanics. I think it's too bad that Everett's very simple thesis (which is just that the linear quantal equations of motion are always exactly right) has come to be called that at all, because that name has sometimes encouraged a false impression that there are supposed to be more physical universes around after a measurement than there were before it. It might have been better to call what Everett came up with a "many-points-of-view" interpretation of quantum mechanics, or something like that, because it is surely true of Everett's picture (as it is in all other pictures of quantum theory that I know about) that there is always exactly one physical universe. However, the rules of Everett's game, which he insists we play to the very end, require that *every* one of the physical systems of which that universe is composed—including cats and measuring instruments and my friend's brain and my *own* brain—can be, and often *are*, in those bizarre superpositions. The various elements of such a superposition, in the case of brains, correspond to a variety of mutually exclusive points of *view* about the world, as it were, all of which are simultaneously associated with one and the same physical observer.

Needless to say, in some given physical situation, different observers may be associated with different numbers of such points of view (they may, that is, inhabit different numbers of Everett worlds). Suppose, for example, that we add a second friend (Friend #2) to the little story we just told. Suppose at the end of that story, when the state of the composite system consisting of p and of the measuring device for y and of Friend #1 is $[\alpha\rangle$, that Friend #2 measures A, where A is a maximal observable of that composite system such that $A[\alpha\rangle = \alpha[\alpha\rangle$. Friend #2 carries out that measurement by means of an A-measuring device (which, according to the rules of the game, can always be constructed) which interacts with that composite system and which Friend #2 subsequently looks at to ascertain the result of the measurement. When that's all done (since the result of this measurement will with certainty be $A = \alpha$), things will look like this:

$$[\beta\rangle = [\text{Believes that } A = \alpha\rangle_{\text{Friend \#2}}$$
$$\cdot [\text{Show that } A = \alpha\rangle_{A\text{-Measuring Device}} \cdot [\alpha\rangle.$$

In this state, Friend #1 inhabits two Everett worlds (the world in which $\sigma_y = +\frac{1}{2}$ and the world in which $\sigma_y = -\frac{1}{2}$, whereas Friend #2 inhabits only one (the world in which $A = \alpha$), which by itself encompasses the entire state $[\beta\rangle$. Moreover, in his single world, Friend #2 possesses something like a photograph of the two worlds which Friend #1 simultaneously inhabits (he possesses, that is, a recording in his measuring device of the fact that $A = \alpha$). By means of his measurement of A, Friend #2 directly *sees* the full superposition of Friend #1's brain states; and indeed, he can even specify the relative sign between those states.

Nothing ought to be very surprising in this, and indeed, it was all very well known to Everett and his readers. So far as Friend #2 is concerned, after all, Friend #1, whatever else he may be, is a physical system out there in the external world; and consequently, according to the rules of our game, Friend #1 ought to be no less susceptible to being *measured* in superpositions than a single subatomic particle. But this need not be the very end of the game. One more move, which is fully in accordance with the rules of the game, is possible—a move that Everett never mentions. Here it is: Suppose, at the end of the slightly longer story we just told, when the state of things is [ ), that Friend #2 shows his photograph of the two Everett worlds that Friend #1 simultaneously inhabits to Friend #1. Suppose, that is, that Friend #1 now looks at the measuring apparatus for A. Well, it's quite trivial to show that the result of such a move will be this:

$$[\gamma\rangle = \Bigg[[\text{Believes that } A = \alpha\rangle_{\text{Friend \#2}}$$

$$\cdot \, [\text{Shows that } A = \alpha\rangle_{A\text{-Measuring Device}}$$

$$\cdot \, \frac{1}{\sqrt{2}} [\text{Believes that } A = \alpha, \text{Believes that } \sigma_y = +\tfrac{1}{2}\rangle_{\text{Friend \#1}}$$

$$\cdot \, [\text{Shows that } \sigma_y = +\tfrac{1}{2}\rangle_{y\text{-Measuring Device}} \cdot [\sigma_y = +\tfrac{1}{2}\rangle_p$$

$$+ \, [\text{Believes that } A = \alpha, \text{Believes that } \sigma_y = \tfrac{1}{2}\rangle_{\text{Friend \#2}}$$

$$\cdot \, [\text{Shows that } \sigma_y = \tfrac{1}{2}\rangle_{y\text{-Measuring Device}} \cdot [\sigma_y = \tfrac{1}{2}\rangle_p \Bigg].$$

This is, in a number of respects, a somewhat extraordinary state of affairs. Let's look at it carefully. To begin with, note that we have brought here an additional degree of freedom of Friend #1's brain explicitly into consideration (I've called this degree of freedom the A-memory of Friend #1). This is the degree of freedom wherein Friend #1 remembers the information in the photograph that Friend #2 has just shown him. Now, what's going on in this state is that Friend #1 still simultaneously inhabits two different and mutually exclusive Everett worlds, one in which $\sigma_y = +\tfrac{1}{2}$ and the other in which $\sigma_y = -\tfrac{1}{2}$; but now, in each of those two worlds separately, Friend #1 knows that $A = \alpha$; he knows, that is, that another world exists; indeed, he has literally seen what amounts to a photograph of that other world!

That's the basic idea, but perhaps I can make it a little clearer now by putting it in a somewhat different language.

Let's imagine telling this story from a purely *external* perspective (from the perspective of some *third* observer, say, who isn't involved in the action at all), without speculating about what it might be like to *be* either Friend #1 or Friend #2, as if they both were merely, say, automata. If we tell it that way, then we'll be able to say what it is that's so interesting about this peculiar state at the end without having to indulge in any talk about multiple worlds.

Here's how that works: Suppose that the state $[\gamma\rangle$ obtains, and that the automaton called Friend #1 is ordered to predict the outcome of an upcoming measurement of σ_y, and suppose that σ_y measurement is carried out, and that the automaton's prediction of that outcome is measured as well; then invariably it will be the case that the outcomes of *those* two measurements (the final measurement of σ_y and the measurement of the automaton's *prediction* about the outcome of that measurement) will coincide. And precisely the same thing applies (when $[\gamma\rangle$ obtains) in an upcoming measurement of A as well (since, when $[\gamma\rangle$ obtains, there is an element of the memory bank of this automaton whose state is correlated to σ_y, and there is also *another* element of the memory bank of this same automaton whose state is, at the same time, correlated to A); and as a matter of fact, precisely the same thing applies even to upcoming measurements of *both A and σ_y* (but note that in this case the *order* in which those two measurements are carried out will be important).

This automaton, then, when $[\gamma\rangle$ obtains, *knows* (in whatever sense it may be appropriate to speak of automata *knowing* things), accurately and simultaneously, the values of both A and σ_y, even though those two observables don't commute. What this means (*leaving aside*, as I said, all of the foggy questions about what it might be like from the perspective of the *automaton*, which is what originally drove us to the talk about multiple worlds) is that this automaton, in this state, is in a position to predict, correctly, without peeking, the outcomes of upcoming measurements of either A or σ_y or both, even though A and σ_y are, according to the familiar dogma about measurement theory, incompatible.

Moreover, no automaton in the world other than this one (no *observer* in the world other than Friend #1, in science fiction talk) can *ever*, even in *principle*, be in a position to simultaneously predict the outcomes of upcoming measurements of precisely those two observables (even though they can, of course, know either one). The possibility of Friend #1's being able to make such predictions hinges on the fact that A is an observable of (among other things) *Friend #1 himself*. There is a well-defined sense here in which this automaton, this friend, has privileged epistemic access to *itself*.

Let me (by way of finishing up) try to expand on *that* just a little bit.

There is *another* famous attempt to suppose that the linear quantum-mechanical equations of motion are invariably the true equations of motion of the wave function of the entire physical world. This attempt goes back to Bohm, and has recently been championed and further developed by John Bell. It's a hidden-variables theory (it is, more precisely, a *completely deterministic* hidden-variables theory, which exactly reproduces the statistical predictions of non-relativistic quantum mechanics by means of an averaging over the various possible values of those hidden variables), and *it* has the same straightforward sort of realistic interpretation as does, say, classical mechanics. It's well known that there are lots of things in this theory that one ought to be unhappy about (I'm thinking mostly about non-locality here); but let's concentrate, for just a moment, on the fact that such a theory is logically *possible*. Since this theory makes all the same predictions as quantum mechanics does, every one of those predictions, including the ones in our story about quantum-mechanical automata, will necessarily arise in this theory, too.

That produces an odd situation. Remember the two automata in the story (Friend #1 and Friend #2). Suppose that $|\gamma\rangle$ obtains, and suppose that things are set up so that some future act of #1 is to be determined by the results of upcoming measurements of σ_y

and A. On Bohm and Bell's theory, there is, right now, a matter of fact about what that act is going to be, and it follows from what we discovered that automaton #1 can correctly *predict* what that act is going to be, but *not* so for automaton #2, nor for any other one, anywhere in the world. So it turns out that it can arise, in a *completely deterministic* physical theory, that an automaton can in principle be constructed that can ascertain certain of its own acts in advance, even though no other automaton, and no external observer whatever—supposing even that they can measure with infinite delicacy and infinite precision—can ascertain them; and that strikes me as something of a surprise.

Perhaps it deserves to be emphasized that there are no paradoxes here, and no violations of quantum theory from which, after all, it was all derived. We have simply discovered a new move here, a move that entirely accords with the rules of quantum mechanics (if the quantum-mechanical rules are all the rules there are) whereby quantum-mechanical observers can sometimes effectively carry out certain measurements on *themselves.* This move just wasn't anticipated by the *founders* of quantum mechanics, and it happens that when you make a move like this, things begin to look very odd, and the uncertainty relations cease to apply in the long-familiar ways. ❧

Acknowledgment

I'm thankful to Deborah Gieringer for her technical assistance in preparing this paper for publication.

⟩₀⟩

THE QUANTUM-CLASSICAL CORRESPONDENCE IN LIGHT OF CLASSICAL BELL'S AND QUANTUM TSIRELSON'S INEQUALITIES

L. A. Khalfin, Université Libre de Bruxelles and
Steklov Mathematical Institute of the Academy of Sciences

We study the well-known problem of the quantum–classical correspon-
dence, or the problem of the reconstruction of the classical world within
quantum theory, which is a basic fundamental dynamical theory. In
connection with this problem, we also study the fundamental problem
of the foundation of statistical physics.

"I do not believe in micro- and macro-laws, but only in (structured)
laws of general validity." —A. Einstein

Fast progress in experimental techniques supports more
and more thorough examinations of the applicability of
quantum theory far beyond the range of phenomena from
which quantum theory arose. For all that, no restrictions
in principle are revealed for its applicability and none
inherently of classical physical systems. However, according to
the Copenhagen interpretation,[1] the fundament of quantum
theory is classical ideas (the classical world) taken equally with
quantum ideas rather than being deduced from the latter.
The Copenhagen interpretation stipulates the joint application
of two description modes, the classical and the quantum, to
a physical world which apparently is "connected" in that it
cannot be divided naturally (from the nature of things) into

---

[1]"The Copenhagen interpretation *is* quantum mechanics."—R. Peirles

two realms, well separated with some gap, one being covered by the quantum description and the other by the classical one. The Copenhagen interpretation specifies an interface between the application domains of the two description modes; being sharp, this interface is conventional and movable. It can be shifted within the limits of a "neutral domain," where both description modes are applicable and conform; this is just the correspondence principle for quantum and classical theories. The "neutral domain" is supposed to be vast enough to contain all physical phenomena immediately perceivable by human observers and also all those which can be successfully described by classical physics and treated as part of the classical, described, macroscopic environment.

The physical content of the correspondence principle is obviously connected with the roominess of the "neutral area." Is that area really so vast as indicated above? Related problems and concepts are discussed in this report.

Recently two disciplines have come into play here, namely the theory of classical and quantum correlations, characterized by Bell's and Tsirelson's inequalities, and the quantum theory of irreversible processes, which originated in the unstable particles (states) quantum decay theory and in fluctuation–dissipation relations. A quantitative criterion for applicability of the classical description was obtained by Dr. B. S. Tsirelson and me in a relevant approximation. It is fulfilled for macroscopic bodies under usual conditions; however, it is possible to design conditions to violate it for the same bodies, allowing for macroscopic quantum phenomena.

I will investigate the following:

1. The algebraic structure of classical Bell's inequalities.

2. The quantum Tsirelson's inequalities.

3. The quasiclassical analogs of Bell's inequalities.

4. The axiomatics of local causal behaviors.

5. The phase transition from the quantum description to the classical description.

6. The reconstruction of the classical world within quantum theory, and a quantitative criterion for the applicability of the classical description within quantum theory. ~709~

7. The problem of the foundations of statistical physics and the quantum decay theory.

8. The macroscopic quantum effects.

In this report I give only the standing points of these problems and our results[2] on these problems. More detailed discussion is given in Khalfin (1978b).

1. The Algebraic Structure of Classical Bell's Inequalities and 2. The Quantum Tsirelson's Inequalities

We now investigate the algebraic structure of classical Bell's and quantum Tsirelson's inequalities in the same manner, and we will see that classical Bell's inequalities are the simplest case from this algebraic point of view.

In the algebra of quantum observables, let us choose a commuting subalgebra $\mathcal{A}_i$. Suppose that observables $A_{ij} \in \mathcal{A}_i$ commute, $[A_{i_1 j_1}, A_{i_2 j_2}] = 0$, if $i_1 \neq i_2$, and in general do not commute, $[A_{i_1 j_1}, A_{i_2 j_2}] \neq 0$, if $i_1 = i_2$. In the classical case $A_{ij} \in \mathcal{A}_i$ are simple c-numbers and commute for all i_1 and i_2. Assume (only for simplicity) that every A_{ij} has a discrete

---

[2]See Caldeira and Leggett (1985), Cirel'son (1980), Diósi (1987, 1988), Fock and Krylov (1947), Joos and Zeh (1985), Joos (1986), and Khalfin (1957a, 1957b, 1965, 1978a).

spectrum; let a_{ijk} be the eigenvalues of A_{ij}, and let P_{ijk} be the spectral projectors. Let us fix some quantum state (either the pure state or density operator case) and let $\langle \dots \rangle$ denote the mean value of the observable in this state. Then

$$P_{j_1 j_2 \cdots}^{k_1 k_2 \cdots} \overset{\text{def}}{=} \langle P_{1 j_1 k_1} P_{2 j_2 k_2} \cdots \rangle \tag{1}$$

is the probability of the coincidence of such events: the measurement of the observable $A_{1 j_1}$ gives the result k_1, that of the observable $A_{2 j_2}$ the result A_2, etc.

PROBLEM

To find the general conditions on the values $P_{j_1 j_2 \cdots}^{k_1 k_2 \cdots}$ which can be expressed in the form of equation (1).

In general this problem has not been solved up to now. In our work we can see some of the not-simple cases. But now we will go to the simplest non-trivial case: $i = 1, 2$; $j = 1, 2$; and $k = 1, 2$. Assume for simplicity that $a_{ijk} = \pm 1 \Rightarrow A_{ij}^2 = 1$.

THEOREM

(Caldeira and Leggett 1985)

$$A_{11} \cdot A_{21} + A_{11} \cdot A_{22} + A_{12} \cdot A_{21} - A_{12} \cdot A_{22} \le 2\sqrt{2} \cdot 1,$$
$$\text{Spec}\{A_{11} \cdot A_{21} + A_{11} \cdot A_{22} + A_{12} \cdot A_{21} - A_{12} \cdot A_{22}\} \tag{2}$$
$$\in [-2\sqrt{2}, 2\sqrt{2}].$$

PROOF

It is possible to prove this result for some more general cases, but a direct and simple elegant proof is as follows:

$$2\sqrt{2} - C = \frac{1}{\sqrt{2}}(A_{11}^2 + A_{12}^2 + A_{21}^2 + A_{22}^2) - C$$
$$= \frac{1}{\sqrt{2}}\left(A_{11} - \frac{A_{21} + A_{22}}{\sqrt{2}}\right)^2 + \frac{1}{\sqrt{2}}\left(A_{12} - \frac{A_{21} - A_{22}}{\sqrt{2}}\right)^2,$$
$$C = A_{11} \cdot A_{21} + A_{11} \cdot A_{22} + A_{12} \cdot A_{21} - A_{12} \cdot A_{22}. \tag{3}$$

For the classical case, in which all of A_{11}, A_{12}, A_{21}, and A_{22} commute (are c-numbers), a trivial inequality for these c-numbers is

$$A_{11} \cdot A_{21} + A_{11} \cdot A_{22} + A_{12} \cdot A_{21} - A_{12} \cdot A_{22} \leq 2 \cdot 1. \quad (4)$$

The inequality in equation (4) gives the algebraic structure of the classical Bell–CHSH inequality for correlation functions,

$$[\langle A_{11} \cdot A_{21} \rangle + \langle A_{11} \cdot A_{22} \rangle + \langle A_{12} \cdot A_{21} \rangle - \langle A_{12} \cdot A_{22} \rangle] \leq 2. \quad (5)$$

The inequality in equation (2) gives the algebraic structure of the quantum Tsirelson's inequality for correlation functions,

$$[\langle A_{11} \cdot A_{21} \rangle + \langle A_{11} \cdot A_{22} \rangle + \langle A_{12} \cdot A_{21} \rangle - \langle A_{12} \cdot A_{22} \rangle] \leq 2\sqrt{2}. \quad (6)$$

The inequalities in equations (5) and (6) are model-independent; that is, they do not depend on the physical mechanism and physical parameters, except the space-time parameters connected with the local causality. We see the principal fundamental gap between the classical Bell's and quantum Tsirelson's inequalities, because the quantum Tsirelson's inequalities do not contain the Planck constant. It is interesting to point out that Tsirelson's quantum inequalities for the general case are the same as for the simplest spin-$\frac{1}{2}$ case.

The class of correlation functions ($\langle \ldots \rangle$), or rather of "behaviors" in the sense discussed in the section "The Axiomatics of Local Causal Behaviors," allowed by the quantum Tsirelson's inequalities is essentially smaller than that allowed by general probabilistic local causality (see "The Axiomatics of Local Causal Behaviors"):

$$[\langle A_{11} \cdot A_{21} \rangle + \langle A_{11} \cdot A_{22} \rangle + \langle A_{12} \cdot A_{21} \rangle - \langle A_{12} \cdot A_{22} \rangle] \leq 4, \quad (7)$$

where $\langle A_{11} \cdot A_{21} \rangle \overset{\text{def}}{=} P_{11}^{11}$ and similarly for the other terms.

In this sense the quantum Tsirelson's inequalities are essentially non-trivial. Therefore their possible violation can be revealed in principle by an experiment. In this case the conception of a local quantum theory would be rejected within the same generality, just as a violation of the Bell's inequalities rejects the conception of a local classical theory. Possible and necessary experiments with $K^0 - \overline{K}^0, D^0 - \overline{D}^0, B^0 - \overline{B}^0$, and their analogs in future high-energy areas were discussed in the author's previous work (Khalfin 1982, 1978b).

The Quasiclassical Analogs of Bell's Inequalities

It is natural to believe that a violation of the classical Bell's inequalities by quantum objects must be small in quasiclassical situations. In this connection we want to obtain inequalities which, holding true for quantum objects, approximate the classical Bell's inequalities in quasiclassical situations. Such inequalities, which of course are model-dependent, were derived in Khalfin and Tsirelson (1985), and we called these the quasiclassical analogs of the classical Bell's inequalities. One example of these inequalities is

$$[\langle A_{11} \cdot A_{21}\rangle + \langle A_{11} \cdot A_{22}\rangle + \langle A_{12} \cdot A_{21}\rangle - \langle A_{12} \cdot A_{22}\rangle] \leq 2 + c\frac{\hbar^2}{\sigma},$$
$$(8)$$

where c is the absolute constant and σ is the (model-dependent) parameter of the quasiclassical approximation ($\sigma^{-1} \to 0$ corresponds to the classical limit; see fig. 1).

The Axiomatics of Local Causal Behaviors

In our previous work, we derived the axiomatics of local causal behaviors, based on the general axioms of the probability theory and the conception of local causality (the relativistic

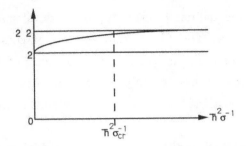

Figure 1. Quasiclassical analogs of Bell's inequalities; phase transition to classical Bell's inequalities.

probability theory). The full classification of all possible behaviors includes (Khalfin 1957b):

a. the stochastic behavior,

b. the deterministic behavior, and

c. the hidden deterministic behavior.

The general stochastic behavior gives us the general inequality in equation (7). The hidden deterministic behavior gives us the classical Bell's inequalities. It is interesting that the so-called dynamical chaos is also a hidden deterministic behavior. All classical stochastic phenomena of the probability theory are hidden deterministic behaviors. And only the quantum behavior gives us the "real" stochastic phenomena.

The Phase Transition from the Quantum Description to the Classical Description

In Khalfin and Tsirelson (1985), we obtained the quasiclassical analogs of Bell's inequalities and some estimates like equation (8). But in Khalfin and Tsirelson (1987) we proved the existence of finite σ_{cr}, and for $\sigma^{-1} \leq \sigma_{cr}^{-1}$ the quasiclassical analogs of Bell's inequalities break down to exactly classical Bell's inequalities. For $\sigma^{-1} \geq \sigma_{cr}^{-1}$ we have the quantum description, the quantum

Tsirelson's inequalities, and the possibility of macroscopic quantum effects. The critical value σ_{cr}^{-1} is of course model-dependent (see the next section). The existence of finite σ_{cr}^{-1} corresponds to the phase transition from the quantum description to the exactly classical (but not quasiclassical) description.

The Reconstruction of the Classical World within Quantum Theory; a Quantitative Criterion for the Applicability of the Classical Description within Quantum Theory

We investigated the problem of the reconstruction of the classical world within quantum theory by using, as in many works,[3] the quantum-dynamical description for our physical objects (systems) and the statistical-physical (thermodynamical) description of the environment. Of course, it will be possible without the logical circle if we investigate the problem of the foundation of statistical physics also within quantum theory (see the next section). In all previous work mentioned, this problem was not investigated.

In these papers the loss of quantum coherence was investigated as the loss of quantum interference and such consideration was necessarily model-dependent. Moreover, in these papers the authors estimated the rate at which quantum coherence is lost and proved only the exponential decrease of this coherence, not the exact breakdown of it.

In our work we investigated the same problem with model-independent tests—the violation of Bell's and Tsirelson's inequalities. Such consideration was true for any of the physical observables. Moreover, in Khalfin and Tsirelson (1987), we proved not

[3]See Khalfin (1983, 1986, 1988), Khalfin and Tsirelson (1985, 1987, 1992), Unruh and Zurek (1989), Wootters and Zurek (1979), and Zurek (1986a, 1986b).

the exponential decrease of the quantum effects, but the disappearance of all quantum effects.

In Zurek (1986b) the speed of delocalization of the quantum correlations was estimated by

$$\Lambda = \hbar^{-2} \cdot k_B \cdot T \cdot \Gamma (sm)^{-2} (\sec)^{-1}, \qquad (9)$$

where k_B is the Boltzmann constant, T is the temperature, and Γ is the friction coefficient. The parts of the wave packet which are divided by interval Δx lose coherence in such a time interval:

$$\Lambda^{-1} \cdot (\Delta x)^{-2} = \frac{\hbar^2}{k_B \cdot T \cdot \Gamma \cdot (\Delta x)^2} (\sec)^1. \qquad (10)$$

In Khalfin (1965) we obtained another estimate on the basis of Bell's and Tsirelson's inequalities:

$$\left(\frac{\hbar m}{k_B \cdot T \cdot \Gamma} \right)^{\frac{1}{2}}, \qquad (11)$$

independent of any information on Δx.

In the same work, we defined an essential new concept—the time of classical factorization (τ_f), that is, the time of loss of quantum effects or quantum description (for a more exact definition, see Khalfin [1965]). For some concrete examples (Khalfin 1965) we obtained quantitative estimates of the corresponding τ_f. For example, for a macroscopic object of length $\simeq 1$ cm and mass $\simeq 1$ g inside a gas with density $\simeq 10^{-26}$ kg/m^3, temperature $\simeq 1$ K, and heat electromagnetic radiation of the same temperature, we obtained the estimate

$$\tau_f \simeq 10^6 \sec. \qquad (12)$$

The estimate in equation (12) gives us the essentially macroscopic time (see the next section).

In Khalfin and Tsirelson (1992), for the estimation of the time of classical factorization τ_f we use a combination of the method in

Khalfin (1965) and some ideas from Diósi's work (Diósi 1987, 1988). Now I give one model example. Let q_1 and q_2 be the coordinates (collective degrees of freedom) of two objects of masses m_1 and m_2 respectively, and let p_1 and p_2 be the corresponding impulses. Let us investigate the quantum dynamics of the motion of these two objects with the Hamiltonian

$$H(q_1, p_1; q_2, p_2) = \frac{1}{2m_1} \cdot p_1^2 + \frac{1}{2m_2} \cdot p_2^2,$$
$$+ \frac{1}{2}k_1 \cdot q_1^2 + \frac{1}{2}k_2 \cdot q_2^2 + k_{12} \cdot q_1 \cdot q_2, \tag{13}$$

and let us add to this interaction the fluctuation forces, which for simplicity we will assume are not correlated for the two objects. The equations of motion, in the language of stochastic differential equations, are

$$dp_1 = -k_1 \cdot q_1 \cdot dt - k_{12} \cdot q_2 \cdot dt + \lambda_1 \cdot db_1, \quad dq_1 = \frac{1}{m_1} \cdot p_1 \, dt,$$

$$dp_2 = -k_2 \cdot q_2 \cdot dt - k_{12} \cdot q_1 \cdot dt + \lambda_2 \cdot db_2, \quad dq_2 = \frac{1}{m_2} \cdot p_2 \, dt, \tag{14}$$

where $b_1(t)$ and $b_2(t)$ are uncorrelated Wiener processes, the derivatives of which are white noise processes,

$$\langle \dot{b}_{1,2}(s) \cdot \dot{b}_{1,2}(t) \rangle = \sigma(s - t), \tag{15}$$

and λ_1 and λ_2 are the intensities of the fluctuating forces. For λ_1 and λ_2, by using the fluctuation–dissipation conditions, the following expression is derived:

$$\lambda_{1,2}^2 = 2 \cdot \Gamma_{1,2} \cdot k_B \cdot T. \tag{16}$$

In Khalfin (1978b) we obtained such necessary and sufficient conditions for reduction to the classical description (the breakdown of quantum Tsirelson's inequalities to classical Bell's inequalities):

$$|k_{12}| \leq \frac{1}{\hbar}\lambda_1 \cdot \lambda_2. \tag{17}$$

This is wonderful, but the condition in equation (17) does not depend on k_1 and k_2. For a simpler case of identical objects we have, from equation (17),

$$
|k_{12}| \leq \frac{1}{\hbar}\lambda^2 = \frac{2}{\hbar}\Gamma \cdot k_B \cdot T = \frac{2\Gamma}{\tau_{\text{therm}}},
$$
$$
\tau_{\text{therm}} \overset{\text{def}}{=} \frac{\hbar}{k_B T}. \tag{18}
$$

For a more general form of interaction with potential energy $U(q_1, q_2)$, the condition in equation (18) is $\quad$ ‑717‑

$$
\left| \frac{\partial^2 U(q_1, q_2)}{\partial q_1 \cdot \partial q_2} \right| \leq \frac{2\Gamma}{\tau_{\text{therm}}}, \tag{19}
$$

which defines the corresponding τ_f. So for times $t > \tau_f$ we see classical (without any quantum interference) dynamics.

The Problem of the Foundation of Statistical Physics; the Quantum Decay Theory

The irreversible decay of unstable particles within reversible quantum theory was the key to the solution of the problem of the foundation of statistical physics within quantum theory.

The irreversible phenomenological decay equation of Rutherford and Soddy,

$$
\frac{d\aleph(t)}{dt} = -\lambda \cdot \aleph(t), \quad \lambda = \text{const}(t) \geq 0, \tag{20}
$$

which was derived before the origin of quantum mechanics, looks like a typical irreversible kinetic equation of statistical physics.

The problem of the foundation of the irreversible phenomenological decay theory is very analogous to the problem of the foundation of statistical physics within the reversible quantum theory. So, the problem is: Is it possible to derive the irreversible phenomenological equation (20) within the reversible quantum theory? If yes, then we must give the method of evaluation of λ from the underlying quantum theory. If no, then we must understand why the phenomenological equation (20) is usually fulfilled with very high accuracy, and we must suggest some problems for which the predictions of the phenomenological theory and the exact underlying quantum theory are quite different.

Let us now investigate the general Cauchy problem of quantum theory for the closed (full Hamiltonian) conservative physical system:

$$H|\Psi(t)\rangle = i\hbar\frac{\partial|\Psi(t)\rangle}{\partial t}, \quad H = \text{const}(t),$$

$$|\Psi_0\rangle = |\Psi(t=0)\rangle, \quad \langle\Psi_0|\Psi_0\rangle = 1,$$

$$H|\varphi_k\rangle = E_k|\varphi_k\rangle, \quad \langle\varphi_k|\varphi_\ell\rangle = \delta_k^\ell,$$

$$H|\varphi_E\rangle = E|\varphi_E\rangle, \quad \langle\varphi_E|\varphi_{E'}\rangle = \delta(E - E'),$$

$$|\Psi_0\rangle = \sum_k c_k|\varphi_k\rangle + \int_{\text{Spec } H} c(E)|\varphi_E\rangle\, dE,$$

$$c_k = \langle\varphi_k|\Psi_0\rangle, \quad c(E) = \langle\varphi_E|\Psi_0\rangle,$$

$$\sum_k |c_k|^2 + \int_{\text{Spec } H} |c(E)|^2 \cdot dE = 1,$$

$$|\Psi(t)\rangle = \exp\left(-\frac{i}{\hbar}Ht\right)|\Psi_0\rangle$$

$$= \sum_k c_k \cdot e^{-\frac{i}{\hbar}E_k t} \cdot |\varphi_k\rangle + \int_{\text{Spec } H} c(E) \cdot e^{-\frac{i}{\hbar}Et} \cdot |\varphi_E\rangle \cdot dE,$$

$$\langle\varphi_k|\Psi(t)\rangle = c_k \cdot e^{-\frac{i}{\hbar}E_k t}, \quad \langle\varphi_E|\Psi(t)\rangle = c(E) \cdot e^{-\frac{i}{\hbar}Et}. \tag{21}$$

From the condition $\langle\Psi_0|\Psi_0\rangle = 1$ it follows that there must exist some self-adjoint operator H_0 (independent of H) for which $|\Psi_0\rangle$ is the eigenvector of the discrete spectrum:

$$H_0|\Psi_0\rangle = E_0^0|\Psi_0\rangle. \tag{22}$$

If we choose different initial vector states $|\Psi_0\rangle$, then H_0 will also be different. The initial vector state $|\Psi_0\rangle$ defines, in addition to and independent of H, the information on the "preparation" or the origin of the investigated physical system. From H and H_0 we can define the interaction part of the Hamiltonian, $H_{\text{int}} = H - H_0$.

Let us define now the decay amplitude $p(t)$:

$$p(t) = \langle \Psi_0 | \Psi(t) \rangle = \langle \Psi_0 | \exp(-\tfrac{i}{\hbar} H t) | \Psi_0 \rangle. \qquad (23)$$

From equation (21) we have the following expression for $p(t)$:

$$p(t) = \sum_k |c_k|^2 \cdot e^{-\frac{i}{\hbar} E_k t} + \int_{\text{Spec } H} |c(E)|^2 \cdot e^{-\frac{i}{\hbar} E t} \cdot dE. \qquad (24)$$

The decay amplitude $p(t)$ is the characteristic function of the probability theory point of view.

DEFINITION

The solution $|\Psi(t)\rangle$ (which was defined by the operator H and the initial vector $|\Psi_0\rangle$, or the operator H_0; see equation (22)) is said to be irreversible if

$$|p(t)| \xrightarrow[t \to \infty]{} 0. \qquad (25)$$

THEOREM

(Fock and Krylov 1947)

For irreversibility of the solution $|\Psi(t)\rangle$ it is necessary and sufficient that

1. H have absolutely continuous spectrum ($c(E) \not\equiv 0$);

2. the contribution of the possible discrete spectrum of H to the initial vector state $|\Psi_0\rangle$ be zero ($c_k \equiv 0 \; \forall k$).

If some $c_k \neq 0$ we have the quantum version of the Poincaré recurrence theorem.

Non-trivial in the proof of this theorem is the necessity, which was based on the famous theorem of S. N. Bernstein from probability theory. From the Fock–Krylov theorem follows the spontaneous breakdown of t-invariance for some solutions of the t-invariant (reversible) quantum theory if the Hamiltonian H has absolutely continuous spectrum.

An essential role in quantum decay theory is played by the spectral principle—the existence of the vacuum state:

$$\text{Spec } H \geq 0. \tag{26}$$

From equation (26) and the Paley–Wiener theorem follows the necessity (Khalfin 1957b, 1957a):

$$\left| \int_{-\infty}^{\infty} \frac{\log |p(t)|}{1 + t^2} \, dt \right| < \infty. \tag{27}$$

Directly from equation (27) follows the non-exponentiality (non-Markovian property) of the decay law. The additional non-exponential term, as was proved in my previous work (Khalfin 1957b, 1957a), is the analytic function of the $\text{Re}\, t > 0$ half-plane and, for this reason, cannot be zero in any interval of time t.

The usual energy distribution $\omega(E) = |c(E)|^2$ satisfies

$$\omega(E) = \xi(E) \cdot \left[(E - E_0)^2 + \Gamma^2 \right]^{-1} \tag{28}$$

where $\xi(E)$ is the continuous "preparation function." From this we have for $t > \hbar / E_0$ the decomposition

$$p(t) \simeq \exp\left(-iE_0 \frac{t}{\hbar} - \frac{\Gamma}{\hbar} |t| \right) - i\frac{\Gamma}{\pi} \cdot \frac{\xi(0) \cdot \hbar}{(E_0^2 + \Gamma^2)} \cdot \frac{1}{t} + o\left(\left(\frac{t}{\hbar} \right)^{-2} \right). \tag{29}$$

The exponential (main term for t of order $(\hbar/\Gamma)^{-1}$) does not depend on the "preparation function," and the non-exponential term for $\Gamma \ll E_0$ for these times is very small.

If $t \to \infty$ and $\Gamma \to 0$ while Γt remains constant, all non-exponential terms disappear.

Now we investigate the problem of the foundation of statistical physics by using methods which are analogous to methods of the quantum decay theory. From the previous section "The Axiomatics of Local Causal Behaviors" it follows that we must investigate this problem within quantum theory. First of all, we must understand that the problems of statistical physics are only special kinds of a general problem of the quantum theory, and must be defined by some additional structure. These problems are defined by the full Hamiltonian H, the initial vector state $|\Psi_0\rangle$ (or H_0), and an additional self-adjoint operator A:

$$H|\Psi(t)\rangle = i\hbar\frac{\partial|\Psi(t)\rangle}{\partial t}, \quad H = \mathrm{const}(t),$$

$$|\Psi_0\rangle = |\Psi(t=0)\rangle, \quad \langle\Psi_0|\Psi_0\rangle = 1, \quad H_0|\Psi_0\rangle = E_0^0|\Psi_0\rangle.$$

$$A|\xi_k\rangle = a_k|\xi_k\rangle.$$

$$(30)$$

The full information in statistical physics is in the set of probabilities $\{P_k(t)\}$, for all k, where

$$P_k(t) = |P_k(t)|^2, \quad P_k(t) = \langle\xi_k|\Psi(t)|\rangle. \qquad (31)$$

It is essential to point out that the full Hamiltonian H includes all quantum fields which define interactions, but A defines the finite number N of particles in the finite box (see fig. 2) for these particles, but not for the fields! For this reason, the full Hamiltonian H, which includes the quantum fields in infinite space, has an absolutely continuous spectrum, which gives us the dynamical (spontaneous) origin of irreversibility.

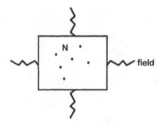

Figure 2. The structure of statistical-physical problems within quantum theory.

Now we can define the usual entropy of statistical physics:

$$S(t) = -\sum_k P_k(t) \log P_k(t),$$

$$0 \leq P_k(t) \leq 1, \quad \sum_k P_k(t) = 1,$$

(32)

for which we must obtain for some special condition the proof of the second law (Boltzmann H-theorem). The usual von Neumann entropy is the dynamical invariant for general problems of quantum theory and has no direct correspondence to the entropy in equation (32) for the problems of statistical physics (this is evident because for von Neumann entropy the second law is not true—this entropy is the dynamical invariant of quantum theory). From equations (30) and (31) it is easy to see that

$$\overline{A}(t) = \langle \Psi(t)|A\Psi(t)\rangle$$

$$= \int_0^\infty \int_0^\infty e^{-\frac{i}{\hbar}Et+\frac{i}{\hbar}E't} \cdot c(E) \cdot c^*(E') \cdot \langle \varphi_{E'}|A\varphi_E\rangle \cdot dE \cdot dE'$$

$$= \sum_k a_k \cdot P_k(t).$$

(33)

If $A = B + G, [B, H] = 0, [G, H] \neq 0, B \neq 0$, and $G \neq 0$, then

$$\langle \varphi_{E'}|A\varphi_E\rangle = b(E') \cdot \delta(E' - E) + g(E', E).$$ (34)

In this case we have

$$\overline{A}(t) = \int_0^\infty b(E) \cdot \omega(E) \cdot d_E$$
$$+ \int_0^\infty \int_0^\infty e^{-\frac{i}{\hbar}Et + \frac{i}{\hbar}E't} \cdot c(E) \cdot c^*(E') \cdot g(E', E) \cdot d_E \cdot d_{E'},$$
$$\overline{A}(\infty) = \int_0^\infty b(E) \cdot \omega(E) \cdot d_E,$$
$$P_k(t) = \int_0^\infty \beta_k(E) \cdot \omega(E) \cdot d_E$$
$$+ \int_0^\infty \int_0^\infty e^{-\frac{i}{\hbar}Et + \frac{i}{\hbar}E't} \cdot c(E) \cdot c^*(E') \cdot \gamma_k(E', E) \cdot d_E \cdot d_{E'},$$
$$\sum_k a_k \cdot \beta_k(E) = b(E),$$
$$\sum_k a_k \cdot \gamma_k(E', E) = g(E', E),$$
$$P_k(\infty) = \int_0^\infty \beta_k(E) \cdot \omega(E) \cdot d_E.$$

$$(35)$$

For the usual $c(E) \sim [E - E_0 + i\Gamma]^{-1}$ we obtain

$$P(t) = P_k(\infty) + \gamma_k \cdot e^{-\frac{2\Gamma}{\hbar}t}$$
$$+ a_0 \cdot e^{-\frac{\Gamma t}{\hbar}} \cdot \frac{\Gamma}{(E_0^2 + \Gamma^2)^2} \cdot \frac{\hbar}{t} \cdot \cos\left(\frac{E_0 t}{\hbar}\right)$$
$$+ b_0 \cdot \frac{\Gamma^2}{(E_0^2 + \Gamma^2)^4} \cdot \frac{\hbar^2}{t^2} + o\left(\left(\frac{t}{\hbar}\right)^{-3}\right),$$

$$(36)$$

and for $\Gamma \ll E_0$ we have

$$A(\infty) = \int_0^\infty b(E) \cdot \omega(E) \cdot d_E \simeq b(E_0 - i\Gamma) \simeq b(E_0). \quad (37)$$

From equations (36) and (37) follows the main statement:[4]

The usual axiomatic statistical physics cannot be the exact theory: the ergodicity and the intermixing,

---

[4]See Diósi (1988), Fock and Krylov (1947), Joos and Zeh (1985), and Khalfin (1957a, 1978a, 1978b).

which are not true, the non-exponential decrease of the correlation functions, and the equilibrium distribution which depends on Γ (the relaxation time). But for usual cases ($\Gamma \ll E_0$) the axiomatic statistical physics is a very good approximation; however, the accuracy of this approximation is not homogeneous for all problems of statistical physics.

BOLTZMANN H-THEOREM

(Khalfin 1978a)

If

$$P_k(t) = P_k(\infty) + \zeta_k(t),$$

$$\frac{d\zeta_k(t)}{dt} = -f(t) \cdot \zeta_k(t), \quad f(t) \geq 0, \tag{38}$$

$$\sum_k \zeta_k(t) \cdot \log P_k(\infty) \geq 0,$$

then

$$\text{a. } \mathcal{S}(t) \leq \mathcal{S}(\infty), \quad \text{b. } \frac{d\mathcal{S}(t)}{dt} \geq 0 \quad \forall t \in [0, \infty). \tag{39}$$

From this theorem it is possible under some conditions on $H, |\Psi_0\rangle$, and A to prove the second law for some big but finite interval of time $t \in [t_1, t_2]$ (see fig. 3). But it also can be proved that for finite small intervals of time $t \in [t_2, t_3]$, in which two non-exponential terms in equation (36) have the same order, for general initial conditions ($|\Psi_0\rangle$) the second law will not be true (see fig. 3). This gives us the first dynamical mechanism for non-special conditions for the origin of order from chaos. This interval of time $t \in [t_2, t_3]$ is an interval of very big times for usual physical systems (God created life [order] on the last day).

The Macroscopic Quantum Effects

As follows from the section "The Phase Transition from the Quantum Description to the Classical Description" (Khalfin

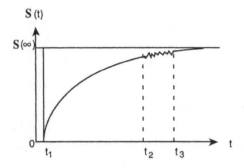

Figure 3. The time dependence of entropy.

1978a; Khalfin and Tsirelson 1992; Zurek 1986a), there exists the time of classical factorization τ_f. For usual macroscopic bodies in usual conditions (T is not very small), τ_f is very small. This is the reason why for usual macroscopic bodies in usual conditions we do not see quantum effects and the description of these bodies for usual times $t > \tau_f$, as was proved, is exactly classical. But for the same macroscopic bodies for another set of not-so-usual conditions (T is sufficiently small and other conditions), τ_f will not be so small, but be macroscopic (as, for example, is shown in eq. (12) of "The Phase Transition from the Quantum Description to the Classical Description," where $\tau_f \simeq 10^6$ seconds). Then for usual macroscopic times $t > \tau_f$, we can see in the dynamics of these macroscopic bodies the macroscopic quantum effects, the quantum correlation, and coherence in the motion. Of course the conditions for such macroscopic τ_f are not very simple, but we hope that in the near future these conditions will be possible for modern experiments to achieve. These macroscopic quantum effects can change our entire classical point of view on the macroscopic world, and may give us the possibility to understand biological phenomena, which as typical quantum phenomena can be characterized by the main property of non-divisibility. 🖎

Acknowledgments

As indicated before, the work reviewed here was done in collaboration with Dr. B. S. Tsirelson. I am indebted to him for interesting collaborations and interesting discussions. My big thanks to the Santa Fe Institute, especially to Prof. J. A. Wheeler and Dr. W. H. Zurek for the invitation to the workshop "Complexity, Entropy and the Physics of Information" (Santa Fe, New Mexico, May 29–June 2, 1989). My big thanks also to the participants of this workshop for interesting discussions. The final version of this report was prepared at the Santa Fe Institute. My big thanks to Dr. George A. Cowan, President of the Santa Fe Institute, for the warm hospitality and the pleasant conditions for scientific work. I thank Prof. T. Toffoli and Dr. W. H. Zurek for improvement of the English version of this report.

REFERENCES

Caldeira, A. O., and A. J. Leggett. 1985. "Influence of Damping on Quantum Interference: An Exactly Soluble Model." *Physical Review A* 31 (2): 1059–66. doi:10.1103/physreva.31.1059.

Cirel'son, B. S. 1980. "Quantum Generalizations of Bell's Inequality." *Letters in Mathematical Physics* 4 (2): 93–100. doi:10.1007/bf00417500.

Diósi, L. 1987. "Exact Solution for Particle Trajectories in Modified Quantum Mechanics." *Physics Letters A* 122 (5): 221–25. doi:10.1016/0375-9601(87)90810-3.

———. 1988. "Continuous Quantum Measurement and Itô Formalism." *Physics Letters A* 129 (8–9): 419–23. doi:10.1016/0375-9601(88)90309-x.

Fock, V. A., and N. S. Krylov. 1947. "On Two Main Interpretations of the Uncertainty Relation for Energy and Time" [in Russian]. *JETP* 17:93–99.

Joos, E. 1986. "Quantum Theory and the Appearance of a Classical World." New Techniques and Ideas in Quantum Measurement Theory, *Annals of the New York Academy of Sciences* 480 (1): 6–13. doi:10.1111/j.1749-6632.1986.tb12404.x.

Joos, E., and H. D. Zeh. 1985. "The Emergence of Classical Properties through Interaction with the Environment." *Zeitschrift für Physik B Condensed Matter* 59 (2): 223–43. doi:10.1007/bf01725541.

Khalfin, L. A. 1957a. "Contribution to the Decay Theory of a Quasi-Stationary State." *JETP* 33:1371–82.

———. 1957b. "On the Theory of the Decay at Quasi-Stationary State" [in Russian]. *Doklady Akad. Nauk SSSR* 115:277–80.

———. 1965. "The Problem of the Foundations of Statistical Physics and the Quantum Theory of Decay." *Doklady Akad. Nauk SSSR* 162 (6): 1273–76. http://mi.mathnet.ru/dan31263.

———. 1978a. "On Boltzmann's H theorem." *Theoretical and Mathematical Physics* 35 (3): 555–58. doi:10.1007/bf01036457.

———. 1978b. "The Statistical Approach to the Ill-Posed Problems of Geophysics." *Zap. Nauchn. Sem. LOMI* 79:67–81.

———. 1982. "Proton Nonstability and the Nonexponentiality of the Decay Law." *Physics Letters B* 112 (3): 223–26. doi:10.1016/0370-2693(82)90967-4.

———. 1983. *Bell's Inequalities, Tsirelson Inequalities, and* $K^0 - \overline{K}^0$, $D^0 - \overline{D}^0$, $B^0 - \overline{B}^0$ *Mesons*. Report on the scientific session of the Nuclear Division of the Academy of Sciences USSR.

———. 1986. "The Problem of the Foundation of the Statistical Physics, the Nonexponentiality of the Asymptotic of the Correlation Functions and the Quantum Theory of Decay." In *Abstracts of the First World Congress of the Bernoulli Society*, edited by Y. V. Prokhorov, II:692. Moscow, Russia: Nauka.

———. 1988. *The Problem of the Foundation of Statistical Physics and the Quantum Decay Theory*. Paper presented at the Stefan Banach International Mathematical Center, September 1988, Warsaw, Poland.

Khalfin, L. A., and B. S. Tsirelson. 1985. "Quantum and Quasi-Classical Analogs of Bell's Inequalities." In *Proceedings of the Symposium on the Foundations of Modern Physics, 1985*, edited by P. Lahti, 441–60. New York, NY: World Scientific.

———. 1987. "A Quantitative Criterion for the Applicability of the Classical Description within the Quantum Theory." In *Proceedings of the Symposium on the Foundations of Modern Physics, 1987*, edited by P. Lahti, 369–401. New York, NY: World Scientific.

———. 1992. "Quantum/Classical Correspondence in the Light of Bell's Inequalities." *Foundations of Physics* 22 (7): 879–948. doi:10.1007/bf01889686.

Unruh, W. G., and W. H. Zurek. 1989. "Reduction of a Wave Packet in Quantum Brownian Motion." *Physical Review D* 40 (4): 1071–94. doi:10.1103/physrevd.40.1071.

Wootters, W. K., and W. H. Zurek. 1979. "Complementarity in the Double-Slit Experiment: Quantum Nonseparability and a Quantitative Statement of Bohr's Principle." *Physical Review D* 19 (2): 473–84. doi:10.1103/physrevd.19.473.

Zurek, W. H. 1986a. "Maxwell's Demon, Szilard's Engine, and Quantum Measurements." In *Frontiers of Nonequilibrium Statistical Physics,* edited by G. T. Moore, 151–61. New York, NY: Plenum. doi:10.1007/978-1-4613-2181-1_11.

———. 1986b. "Reduction of the Wave Packet and Environment-Induced Superselection." New Techniques and Ideas in Quantum Measurement Theory, *Annals of the New York Academy of Sciences* 480 (1): 89–97. doi:10.1111/j.1749-6632.1986.tb12412.x.

———. 1989. Preprint LA-UR-89-225, Los Alamos National Laboratory.

⅄6

SOME PROGRESS IN MEASUREMENT THEORY: THE LOGICAL INTERPRETATION OF QUANTUM MECHANICS

Roland Omnès, Université de Paris-Sud

A few technical advances in the foundations of quantum mechanics, including environment-induced superselection rules, some recent results in semi-classical physics, and Griffiths's consistent histories can be linked together by using a common logical framework to provide a new formulation for the interpretation of the theory which can be favorably compared with the Copenhagen interpretation.

Introduction

Several significant technical advances concerning the interpretation of quantum mechanics have been made more or less recently, mostly during the last decade. I refer particularly to the discovery and study of environment-induced superselection rules,[1] some new general results in semi-classical physics (Ginibre and Velo 1979; Hepp 1974; Omnès 1989a), the distinction to be made between a macroscopic system and a classically behaving one (Leggett 1980, 1984), the possibility to describe a consistent history of a quantum system (Griffiths 1984), and a description of a quantum system by ordinary Boolean logic (Omnès 1988). It turns out that all of these developments can now be joined together to provide a completely new interpretation of quantum mechanics, to be called here the logical interpretation. This name is not coined to mean that the

---

[1]See Caldeira and Leggett (1983, 1984), Feynman and Vernon, Jr. (1963), Hepp and Lieb (1973), and Zurek (1981, 1982).

progress made along the lines of logic is more important than any other advance but to stress the unifying role of logic when bringing these advances together into a consistent theory. The logical interpretation stands upon many fewer axioms than the Copenhagen interpretation and, in fact, upon just a unique universal axiom, and it is not plagued by imprecisely defined words or notions. Its practical consequences, however, coincide mostly with what comes out of the Copenhagen interpretation, except for the removal of some of the disturbing paradoxical features of the latter.

There is no consensus as to what must be considered the most basic difficulties of conventional quantum mechanics. One may use, however, the hindsight provided by recent advances to identify these difficulties with two basic problems, having to do respectively with the status of common sense and the status of empirical facts in quantum mechanics. The first problem comes out of the huge logical gap separating the mathematical framework of the theory (with its Hilbert space and so on) from the ordinary direct physical intuition one has of ordinary physical objects. As will be seen, this is a real problem boiling down to the relation existing between physical reality and its description by mathematics and logic; one will have to make this correspondence clear by stating explicitly how it must be formulated.

The second problem comes from the intrinsically probabilistic character of quantum mechanics: remembering that a theoretical probability can only be checked experimentally by performing a series of trials and noticing that this procedure makes sense only if the result of each individual trial is by itself an undoubtable fact, one sees that quantum mechanics, as an intrinsically probabilistic theory, must, however, provide room for the certainty of the data shown off by a measuring device, i.e., for facts. The solution of this dilemma will involve a proof of the validity of some semi-classical determinism within the framework of quantum mechanics.

A complete interpretation will be obtained by solving these two problems. The general strategy will, however, strongly differ from the Copenhagen approach: classically behaving objects, giving rise to observable facts obeying well determinism and allowing their common-sense description by usual logic, will be interpreted by quantum mechanics and not the other way around. This direct interpretation of what is observed by the most fundamental form of the theory not only is what should be expected from science but also turns out to be both straightforward and fruitful.

General Axioms

The following basic axioms of quantum mechanics will be taken for granted:

Axiom 1 associates a Hilbert space H and an algebra of operators with an individual isolated physical system S or, more properly, with any theoretical model of this system.

Axiom 2 defines dynamics by the Schrödinger equation, using a Hamiltonian H. The corresponding evolution operator will be written as $U(t) = \exp(-2\pi i H t/h)$.

Axiom 3 is technical: the Hilbert space describing two non-interacting systems is the tensor product of their Hilbert spaces, and the total Hamiltonian is the sum of their Hamiltonians.

Von Neumann Predicates

A first step from these very abstract notions toward a more intuitive description of the properties of a system is obtained by using the elementary propositions, or predicates, that were introduced by von Neumann (1932).

First, one should agree about the vocabulary: a self-adjoint operator A will be called as usual an observable (whatever that means in practice), and any real number belonging to the spectrum of A will be called a value of that observable.

Von Neumann considered propositions saying, for instance, that "the position X of a particle is in a volume V." Here the particle is associated with the Hilbert space, X is a well-defined observable, and V is a part of its spectrum, so that everything is well defined in the proposition except the little word "is" or, what amounts to the same, the whole predicate itself. Von Neumann proposed to associate a projector

$$E = \int_V |x\rangle\langle x|\, dx$$

with the predicate to give it a meaning in Hilbert space grammar. More generally, to any set C in the spectrum of an observable A, one can associate a predicate $[A, C]$ meaning "A is in C" and a well-defined projector E. The time-indexed predicate stating that the value of A is in C at time t can be associated with the projector $E(t) = U^{-1}(t)EU(t)$ by taking into account the Schrödinger equation. Conversely, any projector can be used to define a predicate, as can be shown by taking $A = E$ and $C = \{1\}$ in the spectrum of the projector E. One can now define states:

Axiom 4 assumes that the initial state of the system at time zero can be described by a predicate E_0. This kind of description can be shown to represent correctly a preparation process once the theory is complete. A state operator ρ will be defined as the quotient $E_0/\mathrm{Tr}\, E_0$. For instance $\rho = E_0 = |\Psi_0\rangle\langle\Psi_0|$ in the case of a pure state. We shall also freely use, when necessary, the concept of a density matrix.

Histories

As introduced by Griffiths (1984), a history of a quantum system S can be considered as a series of conceptual snapshots describing some possible properties of the system at different times. It will be found later on that a history becomes a true motion picture in the classical limit when the system is macroscopic.

More precisely, let us choose a few ordered times $0 < t_1 < \cdots < t_n$, some observables $A_1, \ldots, A_n$ which are not assumed to commute, and some range of values $C_1, \ldots, C_n$ for each of these observables. A story $[A_1, \ldots, A_n, C_1, \ldots, C_n, t_1, \ldots, t_n]$ is a proposition telling us that at each time t_j $(j = 1, \ldots n)$, A_j has its value in the range C_j.

Griffiths proposed to assign a probability to such a story. We shall write it in the form

$$w = \mathrm{Tr}\big(E_n(t_n) \cdots E_1(t_1)\rho E_1(t_1) \cdots E_n(t_n)\big). \qquad (1)$$

Griffiths used a slightly different expression and relied upon the Copenhagen interpretation to justify it. Here equation (1) will be postulated with no further justification, except to notice that it is "mathematically natural" when using Feynman path summations because a projector $E_j(t_j)$ is associated with a window through which the paths must go at time t_j. It should be stressed that w is just for the time being a mathematical measure associated with the story, having not yet any empirical meaning that could be found by a series of measurements. Quite explicitly, we don't assume that we know right now what a measurement is.

Griffiths noticed that some restrictions must be imposed upon the projectors entering equation (1) in order to satisfy the basic axioms of probability theory and particularly the additivity property of the measures for two disjoint sets. To show what that means, it will be enough to consider the simplest case where time takes only two values t_1 and t_2, denoting by E_1 (E_2 respectively) the projector associated with a set C_1 (C_2 respectively) and by $\overline{E}_1 = I - E_1$ the orthogonal projector. In that case, it can be proved that all the axioms of probability calculus are satisfied by definition in equation (1) if the following *consistency condition* holds:

$$\mathrm{Tr}\big(\big[E_1(t_1), [\rho, \overline{E}_1(t_1)]\big] E_2(t_2)\big) = 0. \qquad (2)$$

One knows how to write down similar necessary and sufficient conditions in the general case. The essential point is that they are completely explicit.

Logical Structure

Griffiths's histories will now be used to describe logically a system in both a rigorous and an intuitive way.

First recall that what logicians call a logic or, more properly, an interpretation of formal logic consists of the following: one defines a *field of propositions* $(a, b, \ldots)$ together with four operations or relations among them, giving a meaning to "*a* or *b*," "*a* and *b*," "not *a*," and "*a* implies *b*," this last relation being denoted by $a \Rightarrow b$ or "if *a*, then *b*." This is enough to do logic rigorously if some twenty or so abstract rules are obeyed by "and," "or," "not," and "if... then." This kind of logic is also called Boolean.

Probability calculus is intimately linked with logic. One can make it clear by choosing, for instance, two times t_1 and t_2 and two observables A_1 and A_2. The spectrum σ_1 of A_1 will be divided into several regions $\{C_{1\alpha}\}$ and similarly for σ_2. An elementary rectangle $C_{1\alpha} \times C_{2\beta}$ in the direct product $\sigma_1 \times \sigma_2$ will be considered as representing a Griffiths's history or what a probabilist would call an elementary event. A union of such sets is what a probabilist calls an event, and here it will be called a *proposition* describing some possible properties of the system.

As usual in set theory, the logical operators "and," "or," and "not" will be associated with an intersection, a union, and the complementation of sets, respectively, so that these three logical rules and the field of propositions or events are well defined.

When a proposition a is associated with a union of two sets a_1 and a_2, each one representing a story, its probability will be defined by

$$w(a) = w(a_1) + w(a_2) \qquad (3)$$

and so on. When the consistency conditions are satisfied, these probabilities are uniquely defined and one can define as usual the conditional probability for a proposition b given some proposition a by

$$w(b \mid a) = \frac{w(a \text{ and } b)}{w(a)}. \qquad (4)$$

Then we shall define "implication" by saying that proposition a implies proposition b ($a \Rightarrow b$) if $w(b \mid a) = 1$. It can be proved that all the axioms of Boolean logic are satisfied by these conventions, as long as the consistency conditions are valid.

We shall also introduce a very important notion here: we shall say that a implies b up to an error ε if $w(b \mid a) > 1 - \varepsilon$. This kind of error in logic is unavoidable when macroscopic objects are concerned. When saying, for instance, that the earth is a satellite of the sun, one must always take into account a small probability ε of the earth leaving the sun and revolving around Sirius by tunnel effect according to quantum mechanics.

You will notice that, even after making sure that the consistency conditions are valid, there remain as many descriptions of a system, or as many logics, as there are choices of the times t_j, the observables A_j, and the different ranges for their values. This multiplicity of consistent logics is nothing but an explicit expression of the complementarity principle.

The calculations that one can perform with these kinds of logic are more or less straightforward, and we shall mention here only one remarkable theorem, albeit a rather simple one: Let us assume that two different logics L_1 and L_2 both contain the same two propositions a and b in their fields of propositions. If $a \Rightarrow b$ in

L_1, then $a \Rightarrow b$ in L_2. This theorem means that no contradiction can ever occur so that the construction can never meet a paradox, in so far as a paradox is a logical conflict.

One can now introduce a unique and universal rule for the interpretation of quantum mechanics, stating how to describe the properties of a physical system in ordinary terms and how to reason about these properties:

Axiom 5: Any description of the properties of a system should be framed into propositions belonging to a consistent logic. Any reasoning concerning them should be the result of an implication or a chain of implications.

From there on, when the word "imply" is used, it will be in the sense of this axiom. The logical construction allows us to give a clear-cut meaning to all the reasonings an experimentalist is bound to make about his apparatuses. In practice, it provides us with an explicit calculus of propositions selecting automatically the propositions making sense and giving the proof of correct reasonings. Two examples will show how this works.

In an interference two-beams experiment, it is possible to introduce the elementary predicates stating that, at some convenient time t_2, a particle is in some region of space where the two beams are recombined. All the predicates corresponding to different regions describe the possible outcomes of the experiment, although one does not know yet how to describe a counting device registering them. They constitute a consistent logic. It is also possible to define a projector expressing that the particle followed the upper beam, but, lo and behold, there is no consistent logic containing this predicate together with the previous predicates describing the outcomes of the experiment. This means that logic precedes measurement. There is no need to invoke an actual measurement to discard as meaningless: the proposition stating that the particle followed the upper beam.

Logic is enough to dispose of it according to the universal rule of interpretation, because there is no consistent logic allowing such a statement.

More positively, one may also consider a particle coming out of an isotropic S-state with a somewhat well-defined velocity. This property can be described by an initial projector E_0. Another projector E_2 corresponds to the predicate stating that the particle has its position within a very small volume δV_2 around a point x_2 at time t_2. Then, one can explicitly choose a time $t_1 < t_2$, construct a volume V_1 that has its center on the way from the source to x_2 and is big enough, and prove the logical implication that "the particle is in δV_2 at time $t_2 \Rightarrow$ the particle is in V_1 at time t_1." So, one can prove in this logical framework that the particle went essentially along a straight trajectory. Similar results hold for the momentum at time t_1. To speak of position and momentum at the same time is also possible, as will be seen later on, but with some restrictions.

Simple as they are, these two examples show that the universal rule of interpretation is able to select meaningful propositions from meaningless ones and also to provide a rationed basis for some common-sense statements which had to be discarded by the Copenhagen interpretation.

Classical Limit

What we have called the universal rule of interpretation makes little avail of what Bohr could have also called a universal rule of interpretation—namely, that the properties of a macroscopic device are described by classical physics. In fact, what he really needed from classical physics was not so much classical dynamics as classical logic where a property can be held to be either true or false, with no probabilistic fuzziness.

Bohr's assumption is not as clear-cut as it once seemed, since Leggett has shown that some macroscopic systems consisting of a superconducting ring that has a Josephson weak link can be in a quantum state (Leggett 1980, 1984). As a consequence, nobody seems to be quite sure anymore what the Copenhagen interpretation really states in this case.

The way out of this puzzle will be found by showing why and when classical physics, i.e., classical dynamics together with classical logic, holds true as a consequence of the universal interpretative rule. This is, of course, a drastic change of viewpoint as compared with the familiar course of physics, since it means that one will try to prove why and when common sense can be applied rather than taking it for granted as a gift of God. In that sense, it is also a scathing attack against philosophical prejudice.

To begin with, one must make explicit what a proposition is in classical physics. One may consider, for instance, giving the position and the momentum of a system within some specified bounds. Such a statement is naturally associated with a cell C in classical phase space (in that case a rectangular cell). Since motion will deform such a cell, it looks reasonable to associate a classical predicate with a more or less arbitrary cell in phase space. It will also be given a meaning as a quantum predicate if one is able to associate a well-defined projector $E(C)$ in Hilbert space with the classical cell C in phase space.

If one remembers that, in semi-classical approximations, each quantum state counts for a cell with volume h^n, n being the number of degrees of freedom, two conditions should obviously be satisfied by the cell C:

1. It must be big enough, i.e., its phase space volume must be much larger than h^n.

2. It should be bulky enough and with a smooth enough boundary to be well tiled by elementary regular cells.

This last condition can be made quite precise, and when both conditions are met and the cell is simply connected, i.e., in one piece with no hole, we shall say that the cell is *regular*.

Now there is a theorem stating that an approximate projector $E(C)$ can be associated with such a regular cell (Hörmander 1979; Omnès 1989a). To be precise, one can define it in terms of coherent (Gaussian) states g_{qp} with average values (q, p) for their position and momentum, setting

$$E(C) = \int_C |g_{qp}\rangle \langle g_{qp}| \, dq \, dp \, h^{-n}. \qquad (5)$$

It is easily found that the trace of $E(C)$ is the semi-classical average number N ($=$ volume of C/h^n) of quantum states in C. In fact, $E(C)$ is not exactly a projector, but one can prove that

$$N^{-1}(C)\text{Tr}\,|E^2(C) - E(C)| = O\big((h/LP)^{1/2}\big), \qquad (6)$$

where L and P are typical dimensions of C along configuration space and momentum space directions. The kind of bound on the trace of an absolute value operator seen in equation (6) is exactly what is needed to obtain classical logic from quantum logics. Using $E(C)$ or a true projector near enough to it, one is therefore able to state a classical property as a quantum predicate. This kind of theorem relies heavily upon microlocal analysis (Hörmander 1985) and, as such, is non-trivial.

One may extend these kinds of kinematical properties to dynamical properties by giving a quantum logical meaning to the classical history of a system. To do so, given the Hamiltonian H, one must first find the Hamilton function $h(q, p)$ associated with it. The answer is given by what is called in microlocal analysis the Weyl symbol of the operator H, and, in more familiar terms, the relation between H and $h(q, p)$ is exactly the one occurring between a density matrix ρ and the associated Wigner distribution function $f_w(q, p)$ (Wigner 1939; Weyl 1950).

Once the Hamilton function $h(q, p)$ is thus defined, one can write down the classical Hamilton equations and discover the cell C_1 which is the transform of an initial regular cell C_0 by classical motion during a time interval t. Of particular interest is the case where C_1 is also regular, and one will then say that the Hamiltonian (or the motion) is regular for the cell C_0 during the time interval t. It will be seen that regular systems are essentially deterministic, hence their great interest.

Since C_0 and C_1 are both regular, one can associate with them two approximate projectors E_0 and E_1 as given by equation (5), satisfying the condition in equation (6). If E_0 were treated like a state operator, it would evolve according to quantum dynamics to become after a time t_1 the operator

$$E_0(t) = U(t)E_0 U^{-1}(t). \tag{7}$$

Another useful theorem, coming from previous results originated by Hepp (1974) and further developed by Ginibre and Velo (1979) and Hagedorn (1981), is the following one (Omnès 1989a): For a regular system, one has

$$N^1(C_0)\text{Tr}\,|E_0(t) - E_1| = O(\varepsilon). \tag{8}$$

Here ε is a small number depending upon C_0, C_1, and t, expressing both the effect of classical motion and wave packet expansion. In a nutshell, this theorem tells us that quantum dynamics logically coincides with classical dynamics, up to an error of order ε, at least when regular systems are considered.

This theorem can be used to prove several results concerning the classical behavior of a regular system. Considering several times $0 < t_1 < \cdots < t_n$ and an initial regular cell C_0 becoming, successively via classical motion, the regular cells $C_1, \ldots, C_n$, one can use the projectors associated with these cells and their complements to build up several quantum propositions. One can then use equation (8) to prove that the quantum logic containing all these predicates is consistent. Furthermore, if one denotes by $[C_j, t_j]$ the proposition stating that the system is in the cell C_j at time t_j (as characterized by the value 1 for the projector $E(C_j)$), one can prove the implication

$$[C_j, t_j] \Rightarrow [C_k, t_k] \qquad (9)$$

for any pair (j, k) in the set $(1, \ldots, n)$. This implication is valid up to an error ε, which is controlled by the characteristic of the cells and the time t_n as explained above.

~743~

Equation (9) has far-reaching consequences. It tells us that classical logic, when expressing the consequences of classical dynamics for a regular system and regular cells, is valid. Of course, it is only valid up to a possible error ε, as shown by the example of the earth leaving the sun or of a car getting out of a parking lot by tunnel effect. This kind of probability is essentially the meaning of the number ε, and its value is specific for each special case to be considered.

Furthermore, the implications in equation (9) entail that the properties of a regular system show, at least approximately, determinism (since the situation at some time t_j implies the situation at a later time t_k). Such a system can also keep a record or a memory (since the situation at a time t_j implies the situation at an earlier time t_k). It will be convenient to call such a chain of mutually implying classical propositions "potential fact." This name is used because determinism and recording are essential characteristics of facts, but one should not forget that at the present stage the theory is still only talk-talk-talk with no supporting experiments, hence the term "potential," meaning an imaginary possibility.

Since Hagedorn (1981) has shown that wave packet spreading is mainly controlled by quantities known from classical dynamics, the property of regularity can in principle be checked completely within classical dynamics. An obvious counter-example of a system not behaving regularly is provided by a superconducting quantum interference device in a quantum situation described by Leggett (1980, 1984) and investigated by several experimentalists (H. Prance et al. 1986; R. J. Prance et al. 1983; R. J. Prance et al. 1985; Tesche 1986). Another example is given by a K-flow after a time t large enough to allow a strong distortion of cells by mixing, and we shall come back to it later on.

Effective Superselection Rules

The dynamical properties consisting of environment-induced superselection rules are well known. I shall, however, recall them briefly for the sake of completeness: Consider, for instance, a real pendulum, i.e., a ball hanging on a wire. The position of the ball's center of mass can be characterized by an angle Θ. This angle is a typical example of a collective coordinate. The other coordinates describing the atoms and the electrons in the ball and the wire are the microscopic coordinates. Their number N is very large and they are collectively called the environment.

One may start from an initial situation where Θ is given and the velocity is zero. More properly, this can be achieved by a Gaussian state $|\Theta\rangle$ realizing these conditions on the average. It may be convenient to assume that the ball and the wire are initially at zero temperature so that the environment is in its ground state $|0\rangle$. So, the complete description of this initial state is given by

$$|\theta) = |\theta\rangle \otimes |0\rangle. \tag{10}$$

Naively, one would say that the motion of the pendulum will generate deformations of the wire and therefore elastic waves or phonons leading to dissipation. If one compares two clearly different initial situations $|\theta_1\rangle$ and $|\theta_2\rangle$, the amount of dissipation in each case after the same time interval will be different so that the corresponding states of the environment will become practically orthogonal as soon as dissipation takes place.

Consider now the initial state

$$|\Psi\rangle = a_1 |\theta_1\rangle + a_2 |\theta_2\rangle$$

and the density operator $\rho = \Psi)(\Psi|$. The collective density matrix ρ_c, describing only the collective coordinate, will be defined as the partial trace of r over the environment. Putting $|\Psi\rangle = a_1 |\theta_1\rangle + a_2 |\theta_2\rangle$, which is a state of the collective degrees of freedom only, one finds easily that

$$\rho_c(0) = (a_1 |\theta_1\rangle + a_2 |\theta_2\rangle)(a_1^* \langle\theta_1| + a_2^* \langle\theta_2|). \qquad \text{(11)}$$

On the other hand, the orthogonality of environmental states noted previously gives, once some dissipation has taken place,

$$\rho_c(t) = |a_1|^2 |\theta_1'\rangle\langle\theta_1'| + |a_2|^2 |\theta_2'\rangle\langle\theta_2'|, \qquad \text{(12)}$$

the state $|\theta_1'\rangle$ being related to the initial state $|\theta_1\rangle$ in a way that exhibits motion and damping which need not interest us here. The essential point is the diagonal form of $\rho_c(t)$, showing the disappearance of phase relations between the two states or what is called an effective superselection rule (Zurek 1981, 1982). It shows that the corresponding potential facts are well separated (distinct), and the theory of measurement that follows will also show them to be exclusive.

As is well known, these naive arguments can be replaced by serious proofs[2] upon which we shall not elaborate, except for a significant remark.

---

[2]See Caldeira and Leggett (1983, 1984), Feynman and Vernon, Jr. (1963), Hepp and Lieb (1973), and Zurek (1981, 1982).

The objection has been raised that effective superselection rules do not provide a final proof of fact separation for two different reasons (Zurek 1981, 1982):

1. When the collective system is an harmonic oscillator and the environment consists of a bath of harmonic oscillators linearly coupled to it, one can prove the existence of Poincaré recurrences, meaning that the macroscopic pendulum may return quite closely to its initial non-separated situation after a long time.

2. One might use a powerful measuring apparatus to detect a microscopic state of the environment which is described by $\rho(t)$, not $\rho_c(t)$, and shows off a non-diagonal matrix element of $\rho(t)$, so that the diagonal form in equation (12) would only be apparent and somewhat irrelevant for the rigorous foundations of quantum mechanics.

These objections are very serious and cannot be discarded in the Copenhagen approach where classical logic is taken as something absolute, but one can easily get rid of them when taking into account the status of classical logic as being an approximation in the logical interpretation. One can easily compute the probability of occurrence of a Poincaré recurrence or of a supermeasuring device giving a non-trivial result (i.e., measuring something). These probabilities turn out to be much smaller than the limit of the validity of classical logic itself and, as such, it makes no sense to take them into account.

Truth, Falsehood, and Perplexity

There is one thing that this theory does not explain: how a unique result of an individual measurement is selected among all the possible outcomes. This common-sense uniqueness of facts is far from trivial when no foreign information is introduced into the theory: why there should be a unique fact is far from obvious and my own opinion on the subject is still wavering.

I shall surely discard hidden classical variables as providing what philosophers would call an efficient cause, because it would be a pity to reintroduce absolute classical logic at a submicroscopic level when we have been able to get rid of its main troubles at the macroscopic level.

Rather than spending too much time with useless words, I shall only call your attention to a few interesting properties of the present construction. To begin with, the logical structure of quantum mechanics, as defined by the probability in equation (1) for a history, *is not a time-reversal invariant.* This is the main difference between our choice of a probability and the one used by Griffiths. On the other hand, the theorems of measurement theory to be given in the next section depend upon this choice.

Using the property of recordable facts, we can imagine a simple world where all past facts could be recorded. In that case, one can split time, for any given time t, into two parts: a past, where all facts are uniquely given, and a future, where all facts must remain potential and all their possible outcomes are allowed. Let us stress that this statement is not a commonplace triviality expressing what we see around us but a mathematical property of the logical construction or, once again, a proof of common sense. So, the theory is not able to provide a cause for the unique occurrence of a fact, but it is able to make place for this uniqueness. Maybe there is no cause after all and the theory just describes what really is.

To get to these deep (or slippery) questions, one can follow Heisenberg's convention by calling true an *actual* fact (i.e., a unique recorded past fact as opposed to a potential one). However, one may go further by relying upon the non-contradiction theorem mentioned in the "Logical Structure" section and consider a statement as *reliable* when it is the logical consequence of a fact. For instance, when I see as a fact the track of a particle in a bubble chamber, I can assert reliably that it came essentially along a straight line before being detected. This is a simple instance where the somewhat formal present theory is nearer to common sense than the Copenhagen interpretation.

Measurement Theory

Measurement theory now becomes a mere exercise (Omnès 1988). To be specific, we shall only consider here the measurement of an observable A belonging to a physical system Q when the eigenvalues $\{a_n\}$ of A are non-degenerate and discrete and the measurement is of the so-called first kind, preserving these eigenvalues. There is no special difficulty in treating more general cases.

A measuring apparatus M will be used to measure the observable A. It will be convenient to consider a collective variable B of M as the measurement data. One can adapt the theory of facts to the case where there is friction and damping. This allows us to consider as data the final position of a dial on a counter or its digital recording. In that case, the observable B can only take, after an irreversible interaction with the environment lasting a time δ, some values $b_0, b_1, \ldots, b_n, \ldots$ which are the experimental data. Initially, B has the neutral value b_0. It should be stressed that the measuring device is treated here by quantum mechanics but, nevertheless and consistently, data are treated like facts.

It will be assumed that Q and M are initially non-interacting and that, because of some wave packet overlapping, they begin to interact at time t_0 and do not interact anymore after time $t_1 = t_0 + \delta$ when M has registered data.

M will be assumed to be a perfect measuring apparatus of the first kind for the observable A. This property can be made explicit by introducing the evolution operator $S = U(t_0, t_1)$ for the $Q + M$ system: it will be assumed that $S |a_n\rangle_Q \otimes |b_0 r\rangle_M$ (i.e., the effect of the interaction upon the initial state $|a_n\rangle$ and a state of M characterized by the neutral initial marking b_0 and degeneracy indices r) is only a linear superposition of some states $|a_q\rangle_Q |b_m, r'\rangle_M$, where $q = m = n$. This semi-diagonality of the S-matrix is the only ingredient that one needs to completely define a measurement.

~749~

Now the logical game consists of introducing many predicates together with their associated projectors: some of them describe the history of Q before measurement and others the history of Q after measurement; a predicate states the initial value b_0, while other predicates mention possible final data b_n; and finally some predicates enunciate the possible values of A at time t_0 and at time t_1. One also introduces the negation of these predicates, to obtain a field of propositions for the measurement process, altogether forming a logic L.

The first question is to decide whether or not this logic L is consistent. To respond, it is convenient to introduce two logics L_1 and L_2 referring only to the measured system Q: L_1 tells stories of Q before measurement and assumes $A = a_n$ (or not) at time t_0; L_2 begins by an initial statement $E_0 = |a_n\rangle\langle a_n|$ at time t_1 and tells stories of Q after measurement.

One can then prove that L is consistent if and only if L_1 and L_2 are both consistent.

The occurrence of the initial predicate E_0 in L_2 is obviously wave packet reduction. Its precise meaning is the following: one can describe the story of Q after measurement once it again becomes an isolated system, but the data $B = b_n$ force us to take the initial preparation predicate E_0. The basic nature of wave packet reduction turns out to be what logicians call in their own language a *modus ponens*: you use, for instance, a *modus ponens* when you apply a theorem while forgetting how you proved it, discarding the corresponding implications. Similarly, one can discard the past history of Q and the whole history of M, taking into account only the data $B = b_n$ when telling the story of Q after measurement.

One can do this consistently, but it is necessary to use E_0 as the initial predicate. Notice that one might have chosen in mathematics to remember all the proofs of all theorems and in physics to follow the story of every apparatus and every particle that came to interact with Q at one time or another. In that sense, wave packet reduction is not really essential; it is only a very convenient result. Note, however, that were we not to use it, we would have to define the initial state at time $t = -\infty$ and maybe introduce the whole universe in our description. So, in that sense, wave packet reduction is really very useful.

Knowing that the overall logic L is consistent, one can try to prove some of its implications. The most interesting one is the following:

$$[B = b_n, t_1] \Rightarrow [A = a_n, t_0], \qquad (13)$$

or, in words, the result $A = a_n$ of the measurement is a logical consequence of the data $B = b_n$. The nature of this relation between *data* and *result* was left in the shadows by the Copenhagen interpretation, leading to difficulties such as the Einstein–Podolsky–Rosen (EPR) paradox.

Another theorem tells us that, under some trivial restrictions, if one can perform once again a measurement of A after the first measurement giving the result a_n, the second result will also be a_n ("repetitivity").

Finally, one can try to compute the probability of the predicate $[B = b_n, t_1]$ describing the experimental data. Because of the semi-diagonality of the S-matrix, this probability turns out to depend only upon the properties of the Q-system and not at all upon the irrelevant degeneracy indices r which represent the model of the apparatus, its type, its color, or its age. This probability is simply given by

$$w_n = \langle a_n \mid U(t_1)\rho_Q U^{-1}(t_1) \mid a_n \rangle, \qquad (14)$$

i.e., Born's value for the probability of the result $A = a_n$. Using Axiom 3, one can now consider a series of independent experimental trials, give, as undoubtable fact, meaning to the result of each trial, and therefore give an empirical meaning to probability as representing the frequency of a given physical result. The final link between the theory and empirical physics is then contained in a last axiom expressing Born's interpretation of the wave function, i.e., *Axiom 6*. The theoretical probability of an experimental result is equal to its empirical frequency.

~751~

So, finally, one has recovered the main results of the Copenhagen interpretation without several of its limitations and its paradoxes. The exact evaluation of these results as providing perhaps an escape from the difficulties of quantum mechanics will presumably need some time and much discussion, and it would be premature to assert it by now. However, it seems rather clear that the resulting interpretation is objective.

On Information and Entropy

To conclude, I shall now consider two questions more akin to the topics of the present workshop.

The first one has to do with K-flows. More precisely, let us consider a macroscopic system whose collective variables, in their classical version, behave like a K-flow. Configuration space can be assumed to be bounded, and, by restricting the energy, phase space can also be considered to be bounded. We shall assume that the Kolmogorov entropy increases with time like $\exp(|t|/\delta)$.

Because of the mixing properties of K-flows, most regular cells will necessarily become irregular by classical motion after a time $t \gg \delta$. The kind of distorted cells one obtains cannot be associated with a quantum projector so that classical logic which describes classical dynamics is no longer valid on a time interval somewhat larger than δ! One can still define quantum consistent with Griffiths's histories, but they refer to observables so contrived that they are of little interest.

One can, however, proceed via a statistical description of successive measurements. Let us divide phase space into a large but finite number of fixed macroscopic regular cells C_j with projectors E_j. With projectors E_j, we can assume that a series of non-destructive measurements allows us to test in which cell the system is at times $0, \Delta t, 2\Delta t, \ldots$, where $\Delta t \cong \delta$. If the initial state is described by a density matrix ρ, the statistical results of such measurements on independent copies of the system will yield the probabilities at time zero:

$$w_j = \text{Tr}(\rho E_j). \tag{15}$$

The same results would follow from the effective density matrix

$$\rho_{\text{eff}}(0) = \sum_j w_j \frac{E_j}{\text{Tr}\, E_j}. \tag{16}$$

One can then follow the successive measurements by using $\rho_{\text{eff}}(0)$, letting it evolve by $U(t)$ during a time interval Δt where the cells remain regular, computing $w_j(\Delta t)$, and reconstructing $\rho_{\text{eff}}(\Delta t)$ from them by using equation (16). The errors can be estimated and they increase only linearly in time. The following results can then be obtained at the rigorous level of theoretical physics in contrast to mathematical physics (Omnès, unpublished manuscript).

1. The entropy

$$S_{\text{eff}} = -k\, \text{Tr}(\rho_{\text{eff}} \log \rho_{\text{eff}}) \tag{17}$$

increases with time.

2. When time tends to infinity, all the w_js tend to a common limit (equiprobability or microcanonical equilibrium).

3. If a slowly varying observable Ω is defined as having a Weyl symbol $\omega(x,p)$ that is slowly varying upon a typical cell, then the average value of Ω as obtained from $\rho_{\text{eff}}(t)$ differs little from the average obtained from $\rho(t) = U(t)\rho U_{-1}(t)$.

To know whether the entropy S_{eff} is objective or not is not a solved problem, but this possibility seems quite reasonable. In any case, this kind of approach seems to open an interesting new line of investigation concerning quantum K-systems.

Obviously, the Hamilton equations do not make sense rigorously. However, one may try to define a classical distribution by

$$f(x,p) = \sum_j w_j \chi_j(x,p),$$

where $\chi_j(x,p)$ is the characteristic function of the domain C_j. The same procedure, using the classical equations of motion, leads to a Markov process for the new w_js (identical to the old ones for $t = 0$). Then one can show that the classical averages for a slowly varying dynamical variable coincide with the quantum averages, except for small linearly increasing errors. So, classical physics is in fact retrieved but only in a statistical sense.

The EPR Experiment

I consider, with suspicion, the view according to which quantum mechanics is a part of information theory because, at least in the opinion of a non-specialist, a precise field of logical propositions should be available and well defined before one can build a theory of information about it. This means that information theory, however useful it may be, comes in principle second after interpretation has been well founded. In that sense, it has been shown above that the information about a physical result obtained from some measurement data proceeds via a strict logical implication, with a negligible loss of information when the measurement is perfect.

The EPR experiment is interesting from this point of view for two reasons, first because it has led to some puzzling considerations about the transfer of information (Glauber 1986). Furthermore, the non-contradiction theorem makes the logical interpretation highly falsifiable since any unsolvable paradox should kill it, and it is interesting to submit it to this awesome test.

Let us, therefore, consider the EPR experiment, for instance, in the old but clear version where one has just two position operators X_1 and X_2 and conjugate momenta P_1 and P_2. Defining the two commuting operators $X = X_1 - X_2$ and $P = P_1 + P_2$, one considers the wave function

$$\langle x, p \,|\, \Psi \rangle = \delta(x - a)\delta(p) \qquad (18)$$

and one performs a precise simultaneous measurement of the two commuting observables X_1 and P_2. Let us assume that these measurements yield two data D_1 and D_2 as read on the corresponding measuring devices.

One can still play the logical game of measurement theory to investigate the consistency of the process and find out its logical consequences (Omnès 1989b). One easily proves, for instance, the intuitively obvious result

"D_1 and D_2" $\Rightarrow$ "$X_1 = x_1$ and $P_2 = P_2$."

However, the troublesome and questionable implication standing at the root of the EPR paradox,

"D_1 and D_2" $\Rightarrow$ "$X_1 = x_1$ and $P_1 = -\rho_2$," just doesn't work because there is no consistent logic according to which it could make sense. So, if one accepts the universal rule of interpretation, there is no hint of a paradox and, furthermore, there can be no superluminal transfer of information since there is no logic in which such information might be consistently formulated. Remembering that information theory is based upon probability theory, one seems to have been all along fighting about propositions for which no consistent probability exists.

The dissolution of the EPR paradox in the logical approach looks very simple, and one may wonder whether this simplicity is not in some sense as puzzling as the old paradox itself. ❧

~ 755 ~

Acknowledgments

Laboratoire de Physique Théorique et Haustes Energies is associated with Laboratoire associé au CNRS.

REFERENCES

Caldeira, A. O., and A. J. Leggett. 1983. "Quantum Tunnelling in a Dissipative System." *Annals of Physics* 149 (2): 374–456. doi:10.1016/0003-4916(83)90202-6.

———. 1984. "Quantum Tunnelling in a Dissipative System (Erratum)." *Annals of Physics* 153 (2): 445. doi:10.1016/0003-4916(84)90027-7.

Feynman, R. P., and F. L. Vernon, Jr. 1963. "The Theory of a General Quantum System Interacting with a Linear Dissipative System." *Annals of Physics* 24:118–73. doi:10.1016/0003-4916(63)90068-X.

Ginibre, J., and G. Velo. 1979. "The Classical Limit of Scattering Theory for Non-Relativistic Many Boson Systems." *Communications in Mathematical Physics* 66 (1): 37–76. doi:10.1007/BF01197745.

Glauber, R. J. 1986. "Amplifiers, Attenuators, and Schrödinger's Cat." *Annals of the New York Academy of Sciences* 480 (1): 336–72. doi:10.1111/j.1749-6632.1986.tb12437.x.

Griffiths, R. J. 1984. "Consistent Histories and the Interpretation of Quantum Mechanics." *Journal of Statistical Physics* 36 (1–2): 219–72. doi:10.1007/BF01015734.

Hagedorn, G. 1981. "Semi-Classical Quantum Mechanics III. The Large-Order Asymptotics and More General States." *Annals of Physics* 135 (1): 58–70. doi:10.1016/0003-4916(81)90143-3.

Hepp, K. 1974. "The Classical Limit for Quantum Mechanical Correlation Functions." *Communications in Mathematical Physics* 35 (4): 265–77. doi:10.1007/BF01646348.

Hepp, K., and E. H. Lieb. 1973. "Phase Transitions in Reservoir-Driven Open Systems with Applications to Lasers and Superconductors." *Helvetica Physica Acta* 46:573–603.

Hörmander, L. 1979. "On the Asymptotic Distribution of the Eigenvalues of Pseudodifferential Operators in $\mathbb{R}^n$." *Arkiv för Matematik* 17:297–313.

———. 1985. *The Analysis of Differential Operators.* Volumes I–IV. Berlin, Germany: Springer.

Leggett, A. J. 1980. "Macroscopic Quantum Systems and the Quantum Theory of Measurement." *Progress of Theoretical Physics Supplement* 69:80–100. doi:10.1143/PTP.69.80.

———. 1984. "Quantum Tunneling in the Presence of an Arbitrary Linear Dissipation Mechanism." *Physical Review B* 30:1208. https://doi.org/10.1103/PhysRevB.30.1208.

Omnès, R. 1988. "Logical Reformulation of Quantum Mechanics. I, II, III." *Journal of Statistical Physics* 53:893–932, 933–55, 957–75. https://doi.org/10.1007/BF01014230, https://doi.org/10.1007/BF01014231, https://doi.org/10.1007/BF01014232.

———. 1989a. "Logical Reformulation of Quantum Mechanics. IV. Projectors in Semi-Classical Physics." *Journal of Statistical Physics* 57 (1/2): 357–82. doi:10.1007/BF01023649.

———. 1989b. "The Einstein–Podolsky–Rosen Problem: A New Solution." *Physics Letters A* 138 (4–5): 157–59. doi:10.1016/0375-9601(89)90018-2.

———. "From Hilbert Space to Common Sense: The Logical Interpretation of Quantum Mechanics." Unpublished manuscript.

Prance, H., T. P. Spiller, J. E. Mutton, R. J. Prance, T. D. Clark, and R. Nest. 1986. "Quantum Mechanical Flux Band Dynamics of a Superconducting Weak Link Constriction Ring." *Physics Letters A* 115 (3): 125–31. doi:10.1016/0375-9601(86)90038-1.

Prance, R. J., T. D. Clark, J. E. Mutton, H. Prance, T. P. Spiller, and R. Nest. 1985. "Localisation of Pair Charge States in a Superconducting Weak Link Constriction Ring." *Physics Letters A* 107 (3): 133–38. doi:10.1016/0375-9601(85)90731-5.

Prance, R. J., J. E. Mutton, H. Prance, T. D. Clark, A. Widom, and G. Megaloudis. 1983. "First Direct Observation of the Quantum Behaviour of a Truly Macroscopic Object." *Helvetica Physica Acta* 56:789–95.

Tesche, C. D. 1986. "Schrödinger's Cat: A Realization in Superconducting Devices." *Annals of the New York Academy of Sciences* 480:36–50. doi:10.1111/j.1749-6632.1986.tb12408.x.

von Neumann, J. 1932. *Mathematische Grundlagen der Quantenmechanik*. Berlin, Germany: Springer.

Weyl, H. 1950. "Ramifications, Old and New, of the Eigenvalue Problem." *Bulletin of the American Mathematical Society* 56:115–39. doi:10.1090/S0002-9904-1950-09369-0.

Wigner, E. P. 1939. "On the Quantum Correction for Thermodynamic Equilibrium." *Physical Review* 40 (5): 749. doi:10.1103/PhysRev.40.749.

Zurek, W. H. 1981. "Pointer Basis of Quantum Apparatus: Into What Mixture Does the Wave Packet Collapse?" *Physical Review D* 24:1516–25. doi:10.1103/PhysRevD.24.1516.

———. 1982. "Environment-Induced Superselection Rules." *Physical Review D* 26 (8): 1862–80. doi:10.1103/PhysRevD.26.1862.

INDEX

A

abnormal fluctuations 465, 469, 471
absorber theory of radiation 564
action at a distance 563
adaptive systems 429
algorithmic complexity 465–468, 549
algorithmic entropy 605
algorithmic information 549
Anderson localization 474
approximate probabilities 627
arrow of time 599, 600, 614, 616
    quantum mechanical 612, 640, 643
    thermodynamical 597, 600, 624,
      644
Aspect experiment 539, 540, 543, 544
asynchronous computation 404, 405
attractors 421–425

B

baby universes 689, 690
Bayesian probability theory 569, 578
Bell's inequalities 550–552, 576, 605,
    708, 709, 711–715
    classical 708, 709, 711–713, 716
big bang 614
big crunch 614
bistability 423
black holes 601, 616
Boltzmann–Gibbs–Shannon (BGS)
    entropy 523, 527, 530
    individual 526
Born's interpretation of the wave
    function 751
branch dependence 661, 662

branches 623, 645, 646, 650, 661, 663

C

canonical ensemble 527
cantori 474
Casimir effect 578, 591
Casimir/Unruh correlations 611
causality 584
    relativistic 505
cellular automata 403, 405, 410, 412,
    420, 454, 551
    deterministic dynamics of 431
    one-dimensional unidirectional
      431
    universal, 403, 412
chaos 713, 724
classical ideal gas 502, 508, 509
classical limit 735, 739
classical logic 739–741, 743, 746, 747,
    752
classical spacetime 677
classicity 658–660, 665, 666, 668
Clausius principle 527
clock spins 400–403, 407, 408,
    411–413
coarse graining 602, 623, 633–638, 640,
    641, 646, 649, 650, 653, 654,
    657–659, 666, 667
coarse-grained density matrix 656
coarse-grained set of histories 634, 638,
    648, 650, 652, 655, 656
coherence 422
coherent states 680, 682, 741
    in quantum cosmology 685

VOL. I
TABLE OF CONTENTS

— Complexity & Evolution —

To order Volume I
of *Complexity, Entropy & the Physics of Information,*
go to WWW.SFIPRESS.ORG

EDITOR

W. H. Zurek

Los Alamos National Laboratory; Santa Fe Institute

CONTRIBUTORS TO THESE VOLUMES

David Z. Albert, *Columbia University*

J.W. Barrett, *The University, Newcastle-Upon-Tyne, UK*

Charles H. Bennett, *IBM Thomas J. Watson Research Center*

Carlton M. Caves, *University of Southern California*

James P. Crutchfield, *University of California, Berkeley*

P. C. W. Davies, *The University, Newcastle-Upon-Tyne, UK*

Murray Gell-Mann, *California Institute of Technology*

Jonathan J. Halliwell, *University of California, Santa Barbara*

James B. Hartle, *University of California, Santa Barbara*

Tad Hogg, *Xerox Palo Alto Research Center*

E.T. Jaynes, *Washington University*

Stuart A. Kauffman, *University of Pennsylvania; Santa Fe Institute*

L. A. Khalfin, *Steklov Mathematical Institute of theAcademy of Sciences, USSR*

Dilip K. Kondepudi, *Wake Forest University*

Seth Lloyd, *California Institute of Technology*

G. Mahler, *Institut für Theoretische Physik, Universität Stuttgart, FRG*

Norman Margolus, *MIT Laboratory for Computer Science*

NOTE: Affiliations listed here are as of the date of first printing.

V. F. Mukhanov, *Institute for Nuclear Research*, USSR

Roland Omnes, *Laboratoire de Physique Théorique et Hautes Energies, Université de Paris-Sud, France*

M. Hossein Partovi, *California State University, Sacramento*

Asher Peres, *Technion, Israel Institute of Technology*

J. Rissanen, *IBM Almaden Research Center*

O.E. Rössler, *Institute for Physical and Theoretical Chemistry, University of Tübingen*, FRG

Benjamin Schumacher, *Kenyon College*

Shin Takagi, *Tohoku University, Japan*

W. G. Teich, *Institut für Theoretische Physik, Universität Stuttgart*, FRG

Tommaso Toffoli, *MIT Laboratory for Computer Science*

Xiao-Jing Wang, *Center for Studies in Statistical Mechanics, University of Texas, Austin*

John Archibald Wheeler, *Princeton University; University of Texas, Austin*

C. H. Woo, *Center for Theoretical Physics, University of Maryland*

William K. Wootters, *Santa Fe Institute; Center for Nonlinear Studies & Theoretical Division, Los Alamos National Laboratory; Williams College*

Karl Young, *University of California, Santa Cruz*

A. Zee, *University of California, Santa Barbara*

H.D. Zeh, *Institut für Theoretische Physik, Universität Heidelberg*, FRG

W.H. Zurek, *Los Alamos National Laboratory; Santa Fe Institute*

THE EDITOR

WOJCIECH ZUREK is a theoretical physicist based at Los Alamos National Laboratory in New Mexico. He received his MSc from the AGH University of Science and Technology in Krakow, in his native Poland, and his PhD in 1979 from the faculty of physics at the University of Texas at Austin. He remained there as a postdoctoral fellow of Professor John Archibald Wheeler until 1981, when he joined the California Institute of Technology as Tolman Fellow. Zurek arrived in Los Alamos in 1984 as an Oppenheimer Fellow. In 1991, he assumed leadership of Theoretical Astrophysics Group, and, in 1996, was appointed Laboratory Fellow in Theory Division. Zurek served as an external faculty member at the Santa Fe Institute, and has been a visiting professor at the University of California, Santa Barbara, where he co-organized the programs *Quantum Coherence and Decoherence* and *Quantum Computing and Chaos* at UCSB's Institute for Theoretical Physics.

Zurek has received numerous awards and distinctions, including Phi Beta Kappa Visiting Lecturer Award (2004), Alexander von Humboldt Prize (2005), Marian Smoluchowski Medal (2009), Einstein Visiting Professorship (Universität Ulm, 2010-2015), Los Alamos Medal (2014), and the titles of Doctor Honoris Causa of the Jagiellonian University (2019) and AGH University of Science and Technology (2021). His most influential contributions include decoherence theory, no-cloning theorem, and the Kibble-Zurek mechanism.

ABOUT THE SANTA FE INSTITUTE

THE SANTA FE INSTITUTE is the world headquarters for complexity science, operated as an independent, nonprofit research and education center located in Santa Fe, New Mexico. Our researchers endeavor to understand and unify the underlying, shared patterns in complex physical, biological, social, cultural, technological, and even possible astrobiological worlds. Our global research network of scholars spans borders, departments, and disciplines, bringing together curious minds steeped in rigorous logical, mathematical, and computational reasoning. As we reveal the unseen mechanisms and processes that shape these evolving worlds, we seek to use this understanding to promote the well-being of humankind and of life on Earth.

THE SANTA FE INSTITUTE PRESS

The SFI Press endeavors to communicate the best of complexity science and to capture a sense of the diversity, range, breadth, excitement, and ambition of research at the Santa Fe Institute. To provide a distillation of work at the frontiers of complex-systems science across a range of influential and nascent topics. *To change the way we think.*

SEMINAR SERIES
New findings emerging from the Institute's ongoing working groups and research projects, for an audience of interdisciplinary scholars and practitioners.

ARCHIVE SERIES
Fresh editions of classic texts from the complexity canon, spanning the Institute's four decades of advancing the field.

COMPASS SERIES
Provocative, exploratory volumes aiming to build complexity literacy in the humanities, industry, and the curious public.

SCHOLARS SERIES
Affordable and accessible textbooks and monographs disseminating the latest findings in the complex-systems science world.

— ALSO FROM SFI PRESS —
Emerging Syntheses in Science
David Pines, ed.

The Energetics of Computing in Life and Machines
Chris Kempes, David H. Wolpert,
Peter F. Stadler & Joshua A. Grochow, eds.

Ex Machina:
Coevolving Machines and the Origins of the Social Universe
John H. Miller

For additional titles, inquiries, or news about the Press, visit us at
WWW.SFIPRESS.ORG

COLOPHON

The body copy for this book was set in EB Garamond, a typeface designed by Georg Duffner after the Ebenolff-Berner type specimen of 1592. Headings are in Kurier, created by Janusz M. Nowacki, based on typefaces by the Polish typographer Małgorzata Budyta. For footnotes and captions, we have used CMU Bright, a sans serif variant of Computer Modern, created by Donald Knuth for use in TeX, the typesetting program he developed in 1978. Additional type is set in Cochin, a typeface based on the engravings of Nicolas Cochin, for whom the typeface is named

The SFI Press complexity glyphs used throughout this book were designed by Brian Crandall Williams.

SANTA FE INSTITUTE
COMPLEXITY
GLYPHS

ZERO

ONE

TWO

THREE

FOUR

FIVE

SIX

SEVEN

EIGHT

NINE

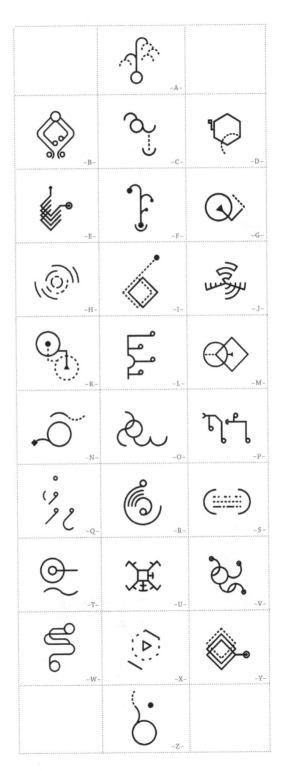

CPSIA information can be obtained
at www.ICGtesting.com
Printed in the USA
LVHW090907240723
753027LV00094B/300/J